I0046451

Small Signal Stability in Power Systems: Concept and Control

Ahmed Abdulsalam Abdulqader

United Scholars Publications, USA

www.unitedscholars.net

Copyright © 2015 Ahmed A. Abdulqader
All rights reserved.
ISBN-13: 978-0692505380
ISBN-10: 0692505385

Dedication

To you, my father, may your soul rest eternally in peace
My mother, for being always everywhere for me
My wonderful wife, beautiful daughter
Brother and Sisters, thank you …

Preface

Synchronous generator stability is an essential issue in the studies of dynamic performance of electric power systems. One important category of stability analysis is the low-frequency oscillations of machine rotor due to disturbances to which the power system is susceptible. These oscillations may sustain and grow in magnitude to cause machine separation if adequate damping is not provided. To enhance system damping, the generating unit is equipped with a power system stabilizer (PSS). Conventional PSS's are widely utilized to damp the low-frequency inertial oscillations. The design of such stabilizers involves finding the set of PSS parameters which yield the best achievable damping response. Several design approaches were proposed over the years and some of them are given in the literature review of this study. A novel genetic-algorithm based optimization approach to design a robust PSS is presented in this study. The proposed approach employs optimization of damping to obtain minimum speed deviation and best possible time-domain transient performance. The single machine infinite bus system is used in this work. Simulations of the linearized system in addition to a simpower model based on the SimPowerSystems® are presented.

The machine speed response has been investigated for both models in the case of Classic-PSS and GA-PSS designs. Their results are compared at different operating conditions, and the comparison shows that the proposed method gives encouraging results against classic method based on system response while being subjected to disturbances which mean that the system maintains its stability during ($\pm$20 %) of perturbations in the load or the system is robustly stable.

Ahmed Abdulsalam Abdulqader

July 2015

Table of Contents

List of Symbols

$\acute{E}$ — machine internal voltage (Volt)

E_B — infinite bus voltage (Volt)

E_{Bd} — infinite bus voltage in the d-axis (Volt)

E_{Bq} — infinite bus voltage in the q-axis (Volt)

e_d — instantaneous value of the stator voltage in the d-axis (Volt)

e_{fd} — field voltage in the d-axis (Volt)

e_q — instantaneous value of the stator voltage in the q-axis (Volt)

f_i — fitness of the i_{th} chromosome

G_{ex} — exciter transfer function

G_{PSS} — PSS transfer function

H — inertia coefficient (MWsec./MVA)

H_T — high tension voltage (Volt)

$\tilde{I}_t$ — phasor value of the stator current (Ampere)

i_d — stator current component in the d-axis (Ampere)

i_{fd} — field circuit current in the d-axis (Ampere)

i_q — stator current component in the q-axis (Ampere)

K_A — exciter gain block value

K_D	damping torque coefficient (pu torque / pu speed)
K_{PSS}	PSS system stabilizer gain block value
K_S	synchronizing torque coefficient
L_T	low tension voltage (Volt)
L_{ads}	saturated value of the mutual inductance in the d-axis (Henry)
L_{aqs}	saturated value of the mutual inductance in the q-axis (Henry)
L_{adu}	unsaturated value of the mutual inductance in the d-axis (Henry)
L_l	leakage inductance (Henry)
max d.	maximum amplitude deviation
P_c	crossover probability
P_e , P_t	electrical machine active power (Watt)
P_m	mutation probability
P_{si}	selection probability
Q_e, Q_t	electrical machine reactive power (Var)
R_E	network resistance (Ohm)
R_a	stator circuit resistance (Ohm)
R_{fd}	field circuit resistance (Ohm)
$\acute{S}$	machine complex power (VA)

$T_1 - T_4$	PSS compensator time constants (Second)
T_F	field circuit time constant (Second)
T_R	terminal voltage sensor time constant (Second)
T_w	PSS wash out time constant (Second)
T_e	electromagnetic torque (N.m)
T_m	mechanical input torque (N.m)
V_t	machine terminal voltage (Volt)
V_{ref}	field reference voltage (Volt)
$v_{PSS}, \Delta v_s$	PSS output voltage (Volt)
v_1	terminal voltage sensor output voltage (Volt)
ω_0	base rotor speed (elec. Rad/s)
ω_d	damped oscillation frequency (Rad/s)
ω_r	rotor speed (elec. Rad/s)
ω_{osc}	electromechanical modes oscillation frequency (Rad/s)
X_E	network reactance (Ohm)
X_T	total system reactance (Ohm)
$\acute{X}_d$	machine transient reactance (Ohm)
Z_{eq}	network equivalent impedance (Ohm)
σ	damping factor

δ	rotor angle (Rad)
ΔT_{ar}	torque component due to armature reaction (N.m)
ΔT_{PSS}	torque component due to PSS(N.m)
$\Delta X_1 - \Delta X_2$	PSS internal voltages
ζ	damping ratio
ψ_{fd}	rotor circuit flux linkage in the d- axis (Wb-t)
Ψ_{ad}	air-gap mutual flux linkage in the d-axis (Wb-t)
Ψ_{aq}	air-gap mutual flux linkage in the q-axis (Wb-t)
Ψ_d	stator flux linkage in the d-axis (Wb-t)
Ψ_q	stator flux linkage in the q-axis (Wb-t)

List of Acronyms

ANN	artificial neural network
CPSS	conventional type PSS
DE	differential evolution
DNA	deoxyribonucleic acid
FLC	fuzzy logic controller
FLPSS	fuzzy logic PSS
GA	genetic algorithm
IAE	integral of absolute error
OHTL	overhead transmission line
PSO	particle swarm optimization
PSS	power system stabilizer
SMIB	single machine infinite bus
SPS	SimPowerSystems®
ST1A	IEEE standard of the static type 1-A excitation system
TCSC	thyristor controlled series compensator
VSC	variable structure control

CHAPTER ONE

INTRODUCTION

Power generation and supply industry has been going through major changes in the last two to three decades. Modern power suppliers are exploring new fields of power engineering for the purpose of providing a dependable source of energy for various consumers. A modern power system must be safe, reliable, uninterrupted, economical and socially acceptable. These criteria compel the power companies to operate their utilities at a higher degree of efficiency which imposes the generation to be at full load most of the time. This point should be driven by the fact that at such levels of loading, the power system must be working within the stability limits.

1.1   Power System Stability

Power system stability may be defined as the ability of a power system to be in a state of operating equilibrium in normal steady-state operating condition and regain an acceptable state of equilibrium after being subjected to a disturbance. This property requires the system to be working in a state of synchronism between all generation facilities and since the main part of generation relies on synchronous machines, it is necessary for adequate system operation to keep all synchronous generators in synchronism with each other in terms of power angle and rotor angle relationships [1]. If one generator or a transmission line is lost, the remaining apparatus should compensate for this loss. But this view has a shortcoming of neglecting the dynamics of the transition from one equilibrium point to another. During this period of transition, it is likely for the system to have increasing low frequency oscillations which may occur in the machine rotor and ultimately lead to system tripping. These problems are related directly to the term "power system stability" [2].

From an economic point of view, it is more efficient to produce electrical power with larger generation units and higher per-unit reactance generation and transmission equipment designs. This imposes more reliance on system controls to compensate for the reduction and offset in stability margins due to these apparatus [3].To gain a better understanding at the stability issue and its sub-categories, a classification should be made which may give a deeper insight to different aspects of stability. Usually, power system stability is classified into three main categories:

1. **Voltage stability**: this is the property of keeping the system steady-state voltage within acceptable range.
2. **Frequency stability**: ability of the system to maintain rated frequency within permissible deviation range.
3. **Angle stability**: which is the ability of the system to remain in synchronism.

These classifications could be given in a more detailed way to draw attention to more subcategories of power system stability, as it is depicted in Fig. (1-1).

Transient angle stability has received a large amount of interest in both research and development fields and much more efforts were concentrated on assuring system operation to meet certain reliability standards and criteria.

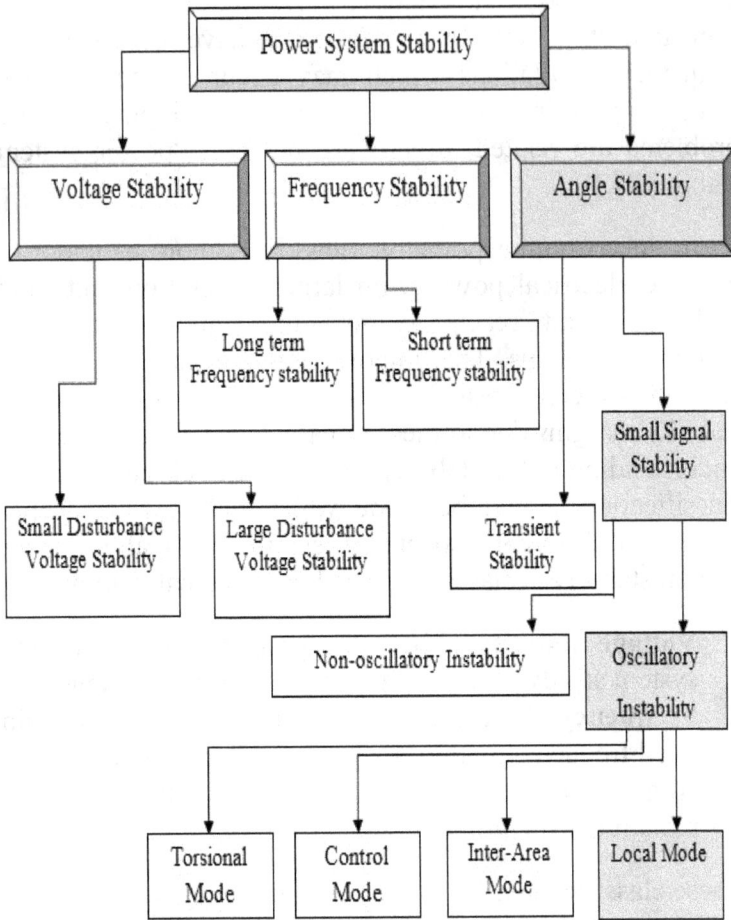

Fig. (1-1) Power system stability classifications

The transient stability is concerned with the dynamics of machine's voltages and currents during and after a large disturbance occurring within the system. But as the duration of the fault is very short, usually for first few seconds after the disturbance, the rotor speed could be considered almost constant. The term "Small signal stability" is defined as the ability of the power system to retain its steady-state characteristics and maintain synchronism after the subjection to a small disturbance. The term "Small disturbance" is used when the equations that describe the resulting system response are linearized for the purpose of analysis. This type of system stability deals with long time scale which the rotor speed will vary and interact with electromechanical changes between the machine and the network to produce electromechanical dynamic effects. The time scale considered with small signal stability is long enough for the associated dynamics to be influenced by the turbine and the generator control systems. Two forms of small signal instability may result [4], [5]:

1. **Rotor angle instability**: which could be in the form of steady increase due to the lack of sufficient synchronizing torque.
2. **Rotor speed instability**: oscillation of increasing amplitude in rotor speed could result due to lack of sufficient damping torque.

1.2 Automatic Voltage Regulators (AVR)

The biggest new advancement in power system controls during the modernizing progression was the introduction and use of modern excitation systems. These fast acting, high gain automatic voltage regulators became essential in enhancing transient stability of the machine following large impacts and also increasing the steady state power limits. But they have a negative effect on small signal stability as these fast excitation changes are not necessarily beneficial in damping the oscillation that follows the first swing and can lead to oscillatory instability in several seconds [6].

1.3   Power System Stabilizers

The excitation control and stabilizing received more attention as an effective way to treat the dynamic or small signal stability phenomena concerned with small perturbations. It was found that adding a stabilizing signal to the excitation could lead to a satisfactory damping results for the low frequency oscillation that occurs due to small perturbations.

These stabilizing signals were added to the voltage error path in the excitation system which would result in providing supplementary damping torque component at the rotor. The device providing the stabilizing signal is known as power system stabilizer or (PSS) [3], [4].

Basically, the PSS is a device which can sense the deviation in the input signal and send out a voltage output signal that is proportional to the input signal but differs with it in phase. The amount of phase difference should compensate the phase lag from the AVR unit and the field circuit in the machine. Thus providing a supplementary damping component that is proportional to the speed changes. The common type of PSS is shown in Fig. (1-2) and it basically consists of a gain block, wash-out filter and a phase compensator. Power system stabilizers are considered to be the most cost-effective electromechanical damping controllers. This is because the necessary amplification power is found within the generator itself and the physical implementation of the device is quite uncomplicated [6], [7]. Power system stabilizers have the ability to enhance power system stability limits where the problem of low frequency oscillation is further apparent.

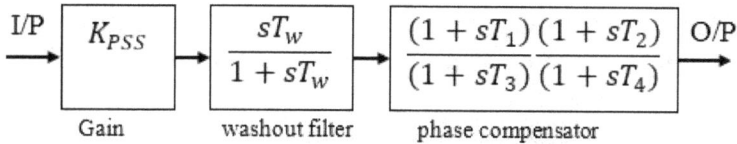

I/P $\rightarrow$	K_{PSS}	$\dfrac{sT_w}{1 + sT_w}$	$\dfrac{(1 + sT_1)(1 + sT_2)}{(1 + sT_3)(1 + sT_4)}$	O/P $\rightarrow$
	Gain	washout filter	phase compensator	

Fig. (1-2) Power system stabilizer block diagram

The most important aspect in the PSS design process is to determine the set of parameters for the phase compensation part which in turn provides the necessary amount of phase in the output signal and then supplies sufficient damping for the system electromechanical modes of oscillation. This should be done without an undesirable effect on other oscillatory forms, such as shaft torsional oscillations, etc. Also, the PSS should recognize that in the case of steady state operation, or when the speed deviation is zero or nearly zero, the voltage controller should be driven by the voltage error only so that the PSS operations do not interfere with the voltage regulation process by the automatic voltage regulator (AVR) [5][6].

1.4 Genetic Algorithms

The roots of genetic algorithms go back to the mid-1950s, where biologists like Barricelli and the computer scientist Fraser began to apply computer-aided simulations in order to gain more insight into genetic processes and the natural evolution and selection. At the beginning of the 1960s, a first achievement was the publication of *Adaptation in Natural and Artificial System* by John H. Holland in 1975 aiming to improve the understanding of natural adaptation process, and to design artificial systems having properties similar to natural systems. As a result of Holland's work, genetic algorithms as a new approach for problem solving were formalized, and finally became widely recognized and popular [28], [29].

Genetic algorithms deal in general with optimizing a certain function. Optimization is concerned with finding either the maxima or the minima point of a function which depends on a number of parameters. The most common type of genetic algorithms has the following major processes or operations:

1. **Initialization** : In which the first population of possible solution is created randomly using a certain type of encoding which converts the original set of solutions (called the pheno-type) into a chain of bits or numbers (called the geno-type) based on the encoding method.

2. **Evaluation**: It is a process which is based on the fitness function of the problem, where the individuals are decoded back to their original values (pheno-type). The fitness of every individual is evaluated and a value is given to each chromosome according to their fitness for the given problem.

3. **Selection**: It is used to choose which chromosomes should survive to form a "mating pool." Chromosomes are chosen based on how fit they are (as indicated by its fitness value) relative to the other members of the

population. More fit individuals end up with more copies of themselves in the mating pool so that they will more significantly affect the formation of the next generation.

4. **Genetic Operators**: In order to produce the next generation, genetic operators manipulate the characters (genes). These operators are :

   4.1 Crossover: In this operation, two random individuals are taken from the mating pool and their genetic information is exchanged at a certain probability.

   4.2 Mutation: This operator causes individual genetic representation to be altered at some point according to some probabilistic rule.

These operations produce the next generation, and the process is repeated iteratively until a desired point is reached based on a certain stopping criterion [8], [9].In this work, the genetic algorithm has been employed to find the best possible set of compensator time constants (T_1, T_2, T_3 and T_4) which should result in a desired stabilization performance over a fairly wide range of operating conditions.

1.5 Literature Reviews

Many researches have been prepared in the last four decades on the subject of PSS design and tuning as it is considered to be one of the most important issues in enhancing the dynamic stability of power systems. The most known type of power system stabilizers is the conventional type PSS (CPSS) with the structure shown in Fig. (1-2). The problem of design which faced the engineers was how to properly set the compensator parameters (i.e. the time constants) yielding in an acceptable system performance.

The concept of supplementary damping torque via excitation was first introduced by DeMello and Concordia (1969) [3], they studied the phenomenon of stability of synchronous machines under small perturbations and provided an insight into effects of excitation systems on small-signal stability to help understand the stabilizing requirements for these systems. Their study paved the way for others to explore the field of oscillations damping via excitation signal modulation.

Over the following years, different design approaches were proposed to present a suitable tuning routine for the CPSS parameters based on traditional and modern design methods. Conventional design based on lead-lag compensator technique was applied by E.V. Larsen and D.A. Swann (1981) [10]. That study showed a general systematic approach to find the set of parameters for CPSS for a given system at a certain operating condition. While this method should provide a maximum amount of damping torque for rotor oscillations at a frequency of oscillation theoretically, other modes of oscillation may suffer from less adequate or even no damping at all.

P. Kundur (1994) [1] presented a broad analysis of the power system models used in small-signal stability studies and conventional PSS (CPSS). It was shown that the proper selection of CPSS parameters should result in satisfactory performance during system upsets.

The gradient procedure in PSS design was used by V. A. Maslennikov et. al. (1997) [11] to optimize the parameters at different operating conditions. Unfortunately, the optimization process requires computations of sensitivity factors at each iteration. This gives rise to heavy computational liability and slow convergence. In addition, the search process is susceptible to be trapped in local minima and the solution obtained will not be optimal.

Fuzzy logic controller (FLC) was employed by M. K. El-Sherbiny et al. (1997) [12] and artificial neural networks (ANN) by M. M. Salem et al. (2000) [13] as power system stabilizers. They are model-free controllers, i.e. they do not require an exact mathematical model of the controlled system. But there was not a practical logical procedure for fuzzy logic PSS (FLPSS) in terms of rules and membership functions which tend to make the design difficult and time consuming task. As with the ANN controllers, they have the capability of adaptation, but it is hard to define the behavior of the network as it is.

Despite the potentials of modern control techniques with different structures, power system operators may still prefer the conventional lead-lag PSS. The reasons behind that might be the ease of on-line tuning and the lack of assurance of the stability related to some adaptive or variable structure techniques (2001) [14].

On the other hand, the advancement in the bio-inspired and evolution techniques like Particle-Swarm Optimization (PSO), Genetic Algorithms (GA) has led to a new approach in solving complex optimization problems. The improvement of using GA techniques is that it is independent of the complexity of the performance index considered. It suffices to specify the objective function and to place finite bounds on the optimized parameters. Y. L. Abdel-Magid et. al. (2003) [15] employed genetic algorithm approach in PSS parameters optimization process. This approach took a multiobjective goal function only, considering the output variable error and taking fixed values for damping factor (σ) and damping ratio (ζ).

N. Al-Musabi et. al. introduced in (2006) [16] a PSS design technique based on Variable Structure Control (VSC) with Particle Swarm Optimization (PSO) approach. They used the PSO technique in designing a VSC for a nonlinear single machine infinite bus system. The drawback for this approach was the need for system state variables to be available for the controller which is based on state feed-back control, and this limits the applicability of the proposed design.

G. Y. Yang et. al. (2008) [17] employed differential evolution (DE) technique in the power system stabilizer design. Differential evolution is recognized as a simple global optimum technique. However, using this technique in multiobjective optimization still raises some issues as it has some difficulties in reaching a desired set of PSS parameters which define the optimum solution. This indicates that multiobjective DE approaches require additional mechanisms to obtain diversity. Also, this technique was proposed originally for problems in which the decision variables are real numbers unlike other optimization techniques for which binary versions exist.

Design schemes based on H_∞ optimization technique were proposed by Jayapal R. (2010) [19] for designing optimum PSS for a single machine infinite bus system. The design process was based on H_∞ loop shaping technique. But it was noticed that the difficulties were associated with the selection of weighting functions of the optimization problem. The additive and/or multiplicative uncertainty representation and composite structure of such stabilizers reduce their applicability.

Other researchers took another path in designing power system stabilizers, relying on modern control techniques. Eslami M. et. al. (2010) [20] proposed a coordination in designing PSS with a TCSC controller where it is done by simultaneously finding the settings for the damping controllers using a sequential quadratic programming technique.

Other design approaches based on genetic algorithms were proposed based on permanent wedge shaped multiobjective function which defines the boundaries for the system eigen values in a fixed wedge area. R. Shivakumar (2010) [21] and Bati A.F.(2010) [22] demonstrated the utilization of GA in finding the optimal set of PSS parameters, some shortcomings were noticed as the problem of design was taken as a single objective function problem and not all PSS parameters were considered adjustable.

Another design method introduced by O. A. Hussain (2011) [23] based on using a methodological approach to design a fuzzy logic controller as a power system stabilizer implemented using MATLAB® simulation environment and on a microcontroller- based software for a single machine infinite bus system.

The present work attempts to solve some of setbacks in previous works, like in [1] and [15]. Genetic algorithm is used to find an optimum set of all PSS time constants by considering an error based multiobjective goal function along with damping optimization technique using the optimum values for damping factor and damping ratio that yield in best damping performance over a wide range of operating conditions. What defines the novelty of this research is the use of a goal function based on speed deviation error as a first objective and damping optimization approach as a second objective.

1.6 The Aim of Study

The aims of this work are to:

- Study the power systems elements' characteristics.

- Develop a state-space model for the power system with these elements (like AVR, etc.).

- Examine different types of Power System Stabilizers (PSS).

- Design a conventional PSS using a traditional method.

- Study various Genetic Algorithms techniques.

- Use Genetic Algorithm to design a robust PSS.

- Evaluate the designed system results.

1.7 Book Outline

Chapter 2 gives in details the procedure for single machine infinite bus (SMIB) system modeling and linearization using system equations. For the purpose of study and analysis, the state-space which describes the linearized system is constructed. The model is investigated for small signal stability and open loop response. PSS structure and functionality are given and classical tuning method based on phase calculation and compensation is used to design a Classic-PSS. The state-space model is modified to include the PSS, and the effect of the PSS on damping is explained.

Chapter 3 presents the theories and elements of genetic algorithms; every component of GA is given in details and the types used in this study are selected and justified. The problem of PSS tuning is treated in this chapter by means of damping and error index optimization using GA. The objective and fitness functions are defined for the proposed GA based optimization software. Results from the optimization process is evaluated against classic design approach using linearized system simulations.

Chapter 4 is where performance assessment is done with the aid of Matlab/SimPowerSystems® blockset. A SMIB model with same parameters and data from chapter 3 are used for design and comparison between Classic- PSS and GA- PSS are different loading conditions and subjected disturbances.

Chapter 5 summarizes and concludes the work presented in this book, draws attention to the important contributions made to the subject of interest and makes suggestions for future works in this area of research and development.

CHAPTER TWO

POWER SYSTEM MODELING

In this chapter, the SMIB system is investigated, the equations that express the stability dynamics are formulated and state-space model of the system concerning the stability aspects also has been prepared. The state-space model was developed and examined to get an insight into the response during system upsets. Traditional method of tuning the PSS was considered to give precise phase compensation. The state-space model was expanded to incorporate the PSS function.

2.1 Mathematical Modeling of the System

The power system considered in this work is a Single Machine connected to an Infinite Bus (SMIB) which is shown in Fig. (2-1).

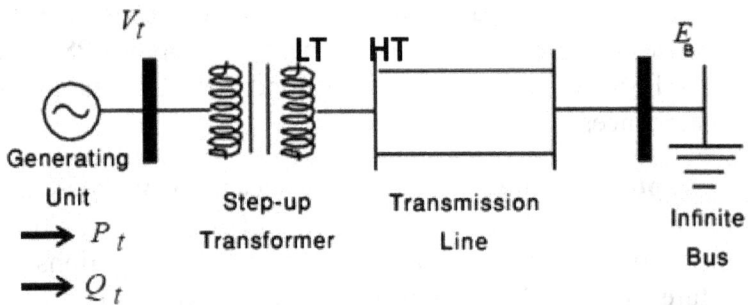

Fig. (2-1) Single Machine Infinite Bus (SMIB) system

The system consists of thermal generating unit connected through a step-up transformer to a two-lines transmission circuit, and then to a power grid that is defined as an infinite-bus [1], [2].

For the purpose of simplicity of analysis, the system of Fig. (2-1) is reduced to the form of Fig. (2-2).by using Thévenin's equivalent of the transmission network. Due to the relatively large size of the system to which the machine is supplying power, electrical dynamics associated with the machine will virtually cause no change in the voltage and frequency of Thévenin's equivalent voltage source E_B which is of constant voltage and frequency and referred to

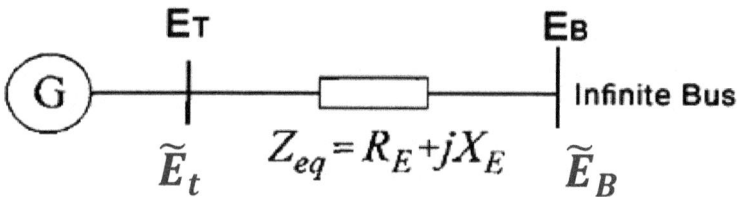

as an *infinite bus*.

Fig. (2-2) Equivalent circuit of SMIB system

For any given system operating condition, the magnitude of the infinite bus voltage E_B remains constant when the machine is perturbed. However, as the steady-state system condition changes, the magnitude of E_B may change representing a changed operating condition of the external network.

The single line diagram of the system is shown in Fig.(2-3) which is based on system representation using equivalent reactances of the transformer and transmission network

(referred   to the LT side of the step-up transformer).
Resistances are assumed to be negligible.

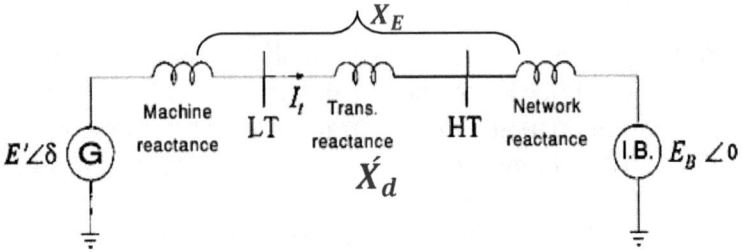

**Fig. (2-3) Single line diagram representation of the
SMIB system**

For the purpose of small-signal stability analysis, linearized
models for a specific operating point are generally
considered to be adequate for representation of the power
system and its various components. In order to develop
such a linearized mathematical model of the power system,
a state-space approach is taken to represent the various
states of the system.

2.1.1   Generator modeling with transient reactance

The generator model in terms of transient electrical
performance, usually referred to as (classical model), is
shown in Fig.(2-4).It is assumed that the flux linking the
main field winding remains constant during the transient
period.

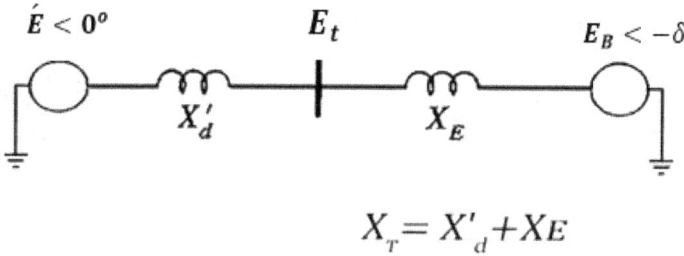

$$X_T = X'_d + X_E$$

Fig. (2.4) SMIB classical model

Here E' is the voltage behind X'_d. Its magnitude is assumed to remain constant and equal the pre-disturbance value. Let δ be the angle by which E' leads the infinite bus voltage E_B. As the rotor oscillates during a disturbance, δ changes.

Taking E' as a reference phasor voltage, then:

$$\tilde{I}_t = \frac{E'\angle 0° - E_B\angle -\delta}{jX_T} = \frac{E' - E_B(\cos\delta - j\sin\delta)}{jX_T} \qquad \dots\dots (2.1)$$

The complex power behind X'_d is given by

$$S' = P + jQ' = \tilde{E}'\tilde{I}^*_t$$

$$= \frac{E'E_B\sin\delta}{X_T} + j\frac{E'(E' - E_B\cos\delta)}{X_T} \qquad \dots\dots (2.2)$$

The air gap power (P_e) is equal to the terminal power (P) neglecting stator resistance. The air-gap torque is equal to the air-gap power in per unit [1], therefore:

$$T_e = P = \frac{E'E_B}{X_T}\sin\delta \qquad \dots\dots (2.3)$$

By linearizing the last equation around an initial operating condition defined by $\delta = \delta_0$, one can get:

$$\Delta T_e = \frac{\partial T_e}{\partial \delta} \Delta \delta = \left[\frac{E' E_B}{X_T} cos\delta_0 \right] (\Delta \delta) \qquad \text{....... (2.4)}$$

The general equations of motion that relate the input mechanical power to the generator with output electric power in per unit are:

$$\frac{d}{dt} \Delta \omega_r = \frac{1}{2H} (T_m - T_e - K_D \Delta \omega_r) \qquad \text{...... (2.5)}$$

$$\frac{d}{dt} \delta = \omega_0 \Delta \omega_r \qquad \text{......(2.6)}$$

Where $\Delta \omega_r$ is the per unit speed deviation, δ is the rotor angle which $\acute{E}$ leads E_B measured in electrical radians, ω_0 is the base rotor electrical speed in radians per second.

Now, by linearizing equation (2.5) and substituting for ΔT_e given by equation (2.4):

$$\frac{d}{dt} \Delta \omega_r = \frac{1}{2H} [T_m - K_S \Delta \delta - K_D \Delta \omega_r] \qquad \text{...... (2.7)}$$

the term $K_D \Delta \omega_r$ is considered as ΔT_D (damping torque coefficient)

Where K_S is the synchronizing torque coefficient given by:

$$K_S = \left[\frac{E' E_B}{X_T} \right] cos\delta_0$$
...... (2.8)

Also by linearizing equation (2.6), we get:

$$\frac{d}{dt} \Delta \delta = \omega_0 \Delta \omega_r$$
...... (2.9)

The two linearized equations, (2.7) and (2.9), could be put into vector-matrix form, as follows:

$$\frac{d}{dt}\begin{bmatrix} \Delta\omega_r \\ \Delta\delta \end{bmatrix} = \begin{bmatrix} -\frac{K_D}{2H} & -\frac{K_S}{2H} \\ \omega_0 & 0 \end{bmatrix}\begin{bmatrix} \Delta\omega_r \\ \Delta\delta \end{bmatrix} + \begin{bmatrix} \frac{1}{2H} \\ 0 \end{bmatrix}\Delta T_m \quad \ldots\ldots \text{(2.10)}$$

which is similar to the general state equation form $\dot{x} = Ax + Bu$. The elements of the system matrix [A] are seen to be dependent on the parameters of the power system K_D, H, K_S and the initial operating condition defined by the values of E' and δ_0 assuming that E_B is constant. The state equation could also be described by the block diagram shown in Fig. (2-5).

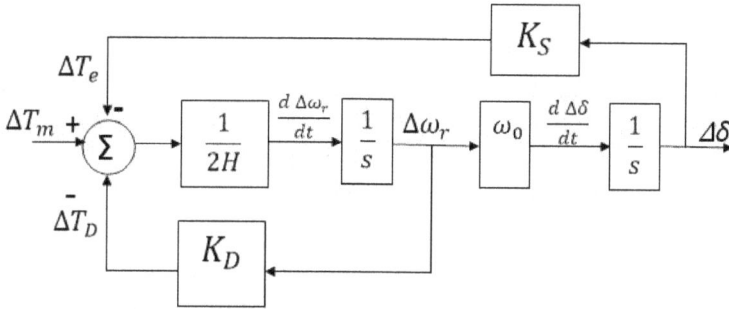

Fig. (2-5) Block diagram representation of state equations in equ. (2.10)

In this system of Fig. (2-5), damping is provided by developing a negative torque proportional to and in phase with speed ω_r. From the block diagram, the following equation (2.11) may be formed:

$$s^2(\Delta\delta) + \frac{K_D}{2H}s(\Delta\delta) + \frac{K_S}{2H}\omega_0(\Delta\delta) = \frac{\omega_0}{2H}\Delta T_m \quad \ldots \text{(2.11)}$$

which gives the following characteristics equation :

$$s^2 + \frac{K_D}{2H}s + \frac{K_S\omega_0}{2H} = 0 \qquad \ldots\text{(2.12)}$$

The damped oscillations frequency is given by $\omega_d = \omega_n\sqrt{1 - \zeta^2}$, where the undamped natural frequency is [3]:

$$\omega_n = \sqrt{K_S \frac{\omega_0}{2H}} \text{rad/s} \qquad \qquad ......(2.13)$$

And the damping ratio is:

$$\zeta = \frac{1}{2}\frac{K_D}{2H\omega_n} = \frac{1}{2}\frac{K_D}{\sqrt{K_S2H\omega_0}} \qquad \qquad ...... (2.14)$$

The instability is approached when the damping ratio becomes less than zero. For normal values of damping ratios and practicable ranges of inertia, impedances and loading values, the frequency of oscillation (ω_d) will be in the range of 0.5 to 2 Hz with the possibility of up to 4 Hz at the high end and 0.1 Hz at the low end of range. At any given oscillating frequency, braking torques developed inside the machine may be split into two components, one in phase with machine rotor angle is called synchronizing torque (T_S), and other in phase with machine rotor speed is termed damping torque (T_D) [3]. Fig. (2-6) illustrates this statement. It can be seen from equations (2.13) and (2.14) that:

1. As the synchronizing torque coefficient K_S increases, the natural frequency ω_n increases.
2. An increase in damping torque coefficient K_D increases the damping ratio ζ.
3. Increasing the inertia constant H decreases both ω_n and ζ.

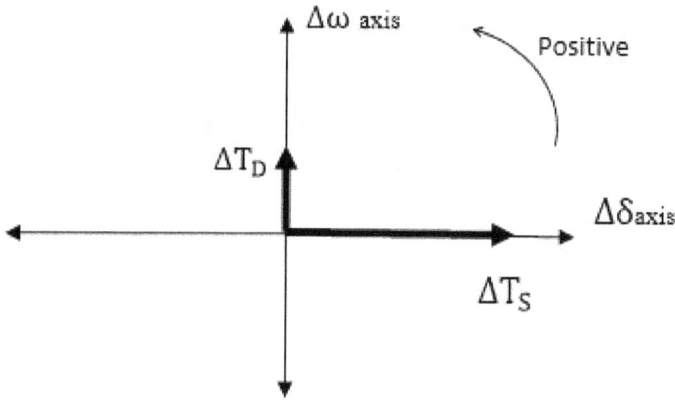

Fig. (2-6) components of braking torques

So as long as K_D is positive, positive damping contribution from the inertial torques is obtained.

2.1.2 Machine field circuit dynamics effect

The system performance is considered here with the inclusion of field flux variation effect. A state-space model of the SMIB system will be developed by reducing the synchronous machine equation and combining them with the network equations. The derivation of the state-space model goes under the following assumptions:

1. The amortisseurs (damper windings) effect is neglected as their effect on small-signal stability is insignificant [1].
2. The network equivalent resistance is small enough to be negligible.

The model takes into account the effects of:

1. Machine field circuit variation.
2. Automatic Voltage Regulator (AVR).
3. Saturation effects on the flux linkages were considered to be insignificant by many models [3], [24], although they have a likely effect on the values of the calculated machine reactances [1].

The system analysis starts with computing of initial steady-state values of system variables. The two-axis synchronous machine representation with a field circuit in the direct axis is considered [25].

The former model of section 2.1.1 may be expanded to include the field circuit dynamics by using the equation:

$$
\left.
\begin{aligned}
\frac{d}{dt}\psi_{fd} &= \omega_0\left(e_{fd} - R_{fd}i_{fd}\right)\\[2mm]
&= \frac{\omega_0 R_{fd}}{L_{adu}}E_{fd} - \omega_0 R_{fd}i_{fd}
\end{aligned}
\right\} \quad \ldots\ldots \text{(2.15)}
$$

where E_{fd} is the exciter output voltage.

Equations (2.7), (2.9) and (2.15) express the dynamics of synchronous machine with $\Delta\omega_r$, $\Delta\delta$ and $\Delta\psi_{fd}$ as the state variables. Nevertheless, the derivatives of these state variables appear in these equations as functions of T_e and i_{fd}, which are neither state variables nor input variables. So T_e and i_{fd} should be expressed in terms of state variables as determined by the machine flux linkage and network equations to develop a complete state-apace model for the system.

Now to develop an expression for i_{fd} and T_e in terms of state variables, two steps will be taken to complete this derivation. First, i_{fd} and T_e will be defined in terms of intermediate variables Ψ_{ad} and Ψ_{aq} which are the air-gap (mutual) flux linkages, and along with i_d and i_q. The second step is to define the current components i_d and i_q in terms of state variables Ψ_{fd} and δ. From Fig. (2-7), the equations describing the stator and rotor flux linkage are:

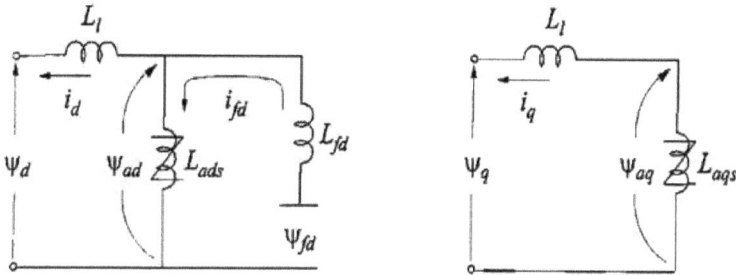

Fig. (2-7) Machine flux linkages and currents equivalent circuit

$$\Psi_d = -L_l i_d + L_{ads}(-i_d + i_{fd})$$

$$= -L_l i_d + \Psi_{ad} \qquad \qquad ......(2.16)$$

$$\Psi_q = -L_l i_q + L_{aqs}(-i_q)$$

$$= -L_l i_q + \Psi_{aq} \qquad \qquad ......(2.17)$$

$$\Psi_{fd} = L_{ads}(-i_d + i_{fd}) + L_{fd} i_{fd}$$

$$= \Psi_{ad} + L_{fd} i_{fd} \qquad \qquad ......(2.18)$$

The terms L_{ads} and L_{aqs} are the saturated values of the mutual inductances. From equ.(2.18), the field current may be defined as :

$$i_{fd} = \frac{\Psi_{fd} - \Psi_{ad}}{L_{fd}} \qquad \qquad ......(2.19)$$

So the term Ψ_{ad} may be expressed as follows:

$$\Psi_{ad} = -L_{ads} i_d + L_{ads} i_{fd}$$

$$= \acute{L}_{ads} \left(-i_d + \frac{\Psi_{fd}}{L_{fd}} \right) \qquad \qquad ......(2.20)$$

And

$$\acute{L}_{ads} = \frac{1}{\frac{1}{L_{ads}} + \frac{1}{L_{fd}}} \qquad \qquad ......(2.21)$$

The q-axis mutual flux is given by:

$$\Psi_{aq} = -L_{aqs} i_q \qquad \qquad ......(2.22)$$

The electrical (air-gap) torque developed is expressed in terms of current components and flux linkages as follows:

$$T_e = \Psi_d i_q - \Psi_q i_d$$

Which could be reduced to the form:

$$= \Psi_{ad} i_q - \Psi_{aq} i_d \qquad \qquad ......(2.23)$$

The stator voltage components equations in the d-q frame could also be formulated in terms of the same intermediate variables as follows:

$$e_d = -R_a i_d - \Psi_q$$

$$= -R_a i_d + (L_l i_q - \Psi_{aq}) \qquad\qquad ......(2.24)$$

$$e_q = -R_a i_q - \Psi_d$$

$$= -R_a i_q + (L_l i_d - \Psi_{ad}) \qquad\qquad ......(2.25)$$

Since there is only one generating unit, this unit as well as the network equations can be expressed in terms of their d-q reference frame. So, referring to Fig.(2-8), the d and q components of the terminal and bus voltages are :

$$\tilde{E}_t = e_d + j e_q \qquad\qquad ......(2.26)$$

$$\tilde{E}_B = E_{Bd} + j E_{Bq} \qquad\qquad ......(2.27)$$

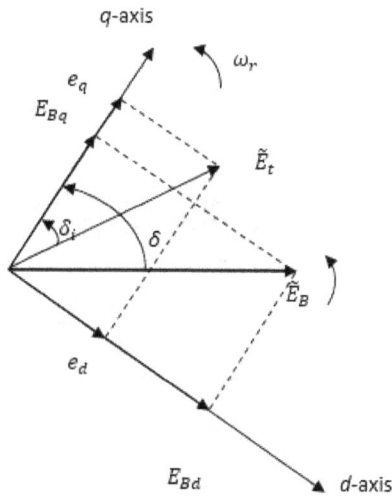

Fig. (2-8) Voltage components phasor diagram in _d-q_ frame

For the network of fig. (2.2), solving the loop voltage equation and applying the network constrains give:

$$\tilde{E}_t = \tilde{E}_B + (R_E + jX_E)\tilde{I}_t \qquad \qquad ......(2.28)$$

$$(e_d + je_q) = (E_{Bd} + jE_{Bq}) + (R_E + jX_E)(i_d + ji_q)$$

Resolving this equation into d-q components gives

$$e_d = R_E i_d - X_E i_q + E_{Bd} \qquad \qquad ......(2.29)$$

$$e_q = R_E i_q - X_E i_d + E_{Bq} \qquad \qquad ......(2.30)$$

where

$$E_{Bd} = E_B sin\delta \quad \text{and} \quad E_{Bq} = E_B cos\delta$$

Using the equations (2.24) and (2.25) to eliminate the terms e_d, e_q and substituting for the expressions of ψ_{ad} and ψ_{aq} from equations (2.20) and (2.22), the following expressions for current components i_d, i_q could be obtained in terms of state variables:

$$i_d = \frac{X_{Tq}\left[\psi_{fd}\left[\frac{L_{ads}}{L_{ads}+L_{fd}}\right]-E_B cos\delta\right]-R_T E_B sin\delta}{D} \qquad ......(2.31)$$

$$i_q = \frac{R_T\left[\psi_{fd}\left[\frac{L_{ads}}{L_{ads}+L_{fd}}\right]-E_B cos\delta\right]+X_{Td} E_B sin\delta}{D} \qquad ......(2.32)$$

$$\left. \begin{array}{l} R_T = R_a + R_E \\[2mm] X_{Tq} = X_E + (L_{aqs}+L_l) = X_E + X_{qs} \\[2mm] X_{Td} = X_E + (L'_{ads}+L_l) = X_E + X'_{ds} \\[2mm] D = R_T^2 + X_{Tq}X_{Td} \end{array} \right\} \qquad ......(2.33)$$

The equations which describe the terms i_d , i_q , Ψ_{ad} , Ψ_{aq} and i_{fd} in (2.31) ,(2.32), (2.16), (2.17) and (2.19) can be used to eliminate i_{fd} and T_e from differential equations (2.7),(2.9) and (2.15) and express them in terms of state variables. But these equations are nonlinear and need to be linearized in order to formulate the state-space model of the system.

After linearizing the equations (2.31), (2.32) using perturbed values expression, one may get:

$$\Delta i_d = m_1 \Delta\delta + m_2 \Delta\Psi_{fd} \qquad\qquad ......(2.34)$$

$$\Delta i_q = n_1 \Delta\delta + n_2 \Delta\Psi_{fd} \qquad\qquad ......(2.35)$$

where

$$m_1 = \frac{E_B(X_{Tq}sin\delta_0 - R_T cos\delta_0)}{D}$$

$$n_1 = \frac{E_B(R_T sin\delta_0 - X_{Td}cos\delta_0)}{D} \qquad ......(2.36)$$

$$m_2 = \frac{X_{Tq}}{D}\frac{L_{ads}}{(L_{ads} + L_{fd})}$$

$$n_2 = \frac{R_T}{D}\frac{L_{ads}}{(L_{ads}+L_{fd})} \qquad\qquad ......(2.37)$$

Now, by linearizing equations (2.20),(2.22) and substituting in the above expressions for Δi_d and Δi_q, one may get :

$$\Delta\psi_{ad} = L'_{ads}\left[-\Delta i_d + \frac{\Delta\psi_{fd}}{L_{fd}}\right]$$

$$= \left[\frac{1}{L_{fd}} - m_2\right]L'_{ads}\Delta\psi_{fd} - m_1 L'_{ads}\Delta\delta \quad ......(2.38)$$

$$\Delta\psi_{aq} = -L_{aqs}\Delta i_q$$

$$= n_2 L_{aqs}\Delta\psi_{fd} - n_1 L_{aqs}\Delta\delta \qquad\qquad \text{......(2.39)}$$

Linearizing the equation (2.19) and substituting for $\Delta\psi_{ad}$ from equation (2.38) gives

$$\Delta i_{fd} = \frac{\Delta\psi_{fd} - \Delta\psi_{ad}}{L_{fd}}$$

$$= \frac{1}{L_{fd}}\left[1 - \frac{L'_{ads}}{L_{fd}} + m_2 L'_{ads}\right]\Delta\psi_{fd} + \frac{1}{L_{fd}}m_1 L'_{ads}\Delta \text{ ...(2.40)}$$

The air-gap torque equation is given as

$$T_e = \psi_{ad}i_q - \psi_{aq}i_a$$
$$\text{......(2.41)}$$

Linearizing this equation(2.41) gives

$$\Delta T_e = \psi_{ad0}\Delta i_q + i_{q0}\Delta\psi_{ad} - \psi_{aq0}\Delta i_d - i_{d0}\Delta\psi_{aq}$$

$$\text{......(2.42)}$$

Given that the variables $\Delta i_d, \Delta i_q, \Delta\psi_{ad}$ and $\Delta\psi_{aq}$ are described from equations (2.34), (2.35), (2.38) and (2.39) respectively, we obtain:

$$\Delta T_e = K_1\Delta\delta + K_2\Delta\psi_{fd} \qquad\qquad \text{......(2.43)}$$

where

$$K_1 = n_1\left(\psi_{ad0} + L_{aqs}i_{d0}\right) - m_1\left(\psi_{aq0} + L'_{ads}i_{q0}\right)$$

$$\text{......(2.44)}$$

$$K_2 = n_2\left(\psi_{ad0} + L_{aqs}i_{d0}\right) - m_2\left(\psi_{aq0} + L'_{ads}i_{q0}\right)$$

$$+\frac{L'_{ads}}{L_{fd}}i_{q0} \qquad\qquad \dots\dots(2.45)$$

Now, using equ. (2.40) and (2.42) along with the linearized form of equations (2.7), (2.9) and (2.15),it is possible to describe the dynamics of the synchronous machine.

$$\begin{bmatrix} \Delta\dot{\omega}_r \\ \Delta\dot{\delta} \\ \Delta\dot{\psi}_{fd} \end{bmatrix} = \begin{bmatrix} a_{11} & a_{12} & a_{13} \\ a_{21} & 0 & 0 \\ 0 & a_{32} & a_{33} \end{bmatrix} \begin{bmatrix} \Delta\omega_r \\ \Delta\delta \\ \Delta\psi_{fd} \end{bmatrix} + \begin{bmatrix} b_{11} & 0 \\ 0 & 0 \\ 0 & b_{32} \end{bmatrix} \begin{bmatrix} \Delta T_m \\ \Delta E_{fd} \end{bmatrix}$$

where

$$a_{11} = -\frac{K_D}{2H} \quad a_{12} = -\frac{K_1}{2H} \quad a_{13} = -\frac{K_2}{2H}$$

$$a_{21} = \omega_0 = 2\pi f_0 a_{32} = \frac{\omega_0 R_{fd}}{L_{fd}}m_1 L'_{ads}$$

$$a_{33} = -\frac{\omega_0 R_{fd}}{L_{fd}}\left[1 - \frac{L'_{ads}}{L_{fd}} + m_2 L'_{ads}\right]$$

$$b_{11} = \frac{1}{2H} \qquad b_{32} = \frac{\omega_0 R_{fd}}{L_{adu}}$$

2.1.3 System state-space modeling with excitation system

The input signal to the excitation system is normally the generator terminal voltage E_T. Therefore, it has to be expressed in terms of state variables $\Delta\omega_r$, $\Delta\delta$ and $\Delta\psi_{fd}$ in order to develop a modified state-space model which includes the effects of the excitation system.

Equation (2.26) shows that:

$$\tilde{E}_t = e_d + je_q$$

By squaring this equation and applying small perturbation, it is linearized:

$$(E_{to} + \Delta E_t)^2 = (e_{do} + \Delta e_d)^2 + (e_{qo} + \Delta e_q)^2$$
$$......(2.46)$$

The above equation could be reduced to the form:

$$E_{to}\Delta E_t = e_{do}\Delta e_d + e_{qo}\Delta e_q$$

Therefore

$$\Delta E_t = \frac{e_{do}}{E_{to}}\Delta e_d + \frac{e_{qo}}{E_{to}}\Delta e_q \qquad ......(2.47)$$

Further derivation of equation (2.47) and substituting the variables Δe_d, Δe_q in terms of state variables would lead to:

$$\Delta E_t = K_5\Delta\delta + K_6\Delta\psi_{fd} \qquad ......(2.48)$$

where

$$K_5 = \frac{e_{do}}{E_{to}}[-R_a m_1 + L_l n_1 + L_{aqs}n_1]$$

$$+ \frac{e_{qo}}{E_{to}}[-R_a n_1 + L_l m_1 + L'_{ads}m_1] \qquad ......(2.49)$$

$$K_6 = \frac{e_{do}}{E_{to}}[-R_a m_2 + L_l n_2 + L_{aqs}n_2]$$

$$+ \frac{e_{qo}}{E_{to}}\left[-R_a n_2 + L_l m_2 + L'_{ads}\left(\frac{1}{L_{fd}}m_2\right)\right] \qquad ......(2.50)$$

The excitation system used here is a static excitation system (type-ST1A). It will be represented by a simplified form that includes necessary elements for modeling. A high exciter gain, without transient gain reduction or derivative feedback is used and also includes the terminal voltage transducer with time constant T_R as shown in Fig. (2-9) [2].

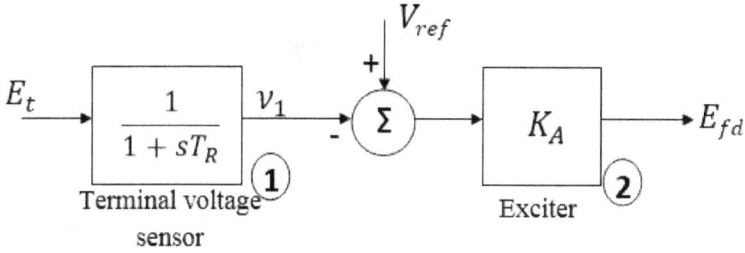

Fig. (2-9) Excitation system type ST1A block diagram

From block (1) of Fig. (2-9) using perturbed value of E_t, we have:

$$\Delta v_1 = \frac{1}{1+sT_R} \Delta E_t \qquad , \text{ Hence}$$

$$s\Delta v_1 = \frac{1}{T_R}(\Delta E_t - \Delta v_1)$$

Now by substituting the term ΔE_t from equation (2.48), we get

$$s\Delta v_1 = \frac{K_5}{T_R}\Delta\delta + \frac{K_6}{T_R}\Delta\psi_{fd} - \frac{1}{T_R}\Delta v_1 \qquad \ldots\ldots(2.51)$$

From block (2) of Fig. (2-9):

$$E_{fd} = K_A(V_{ref} - v_1)$$

Also by applying perturbation to the values, we have

$$\Delta E_{fd} = K_A(-\Delta v_1) \qquad \ldots\ldots(2.52)$$

Now, returning to the field circuit equation (2.15) developed in the previous section, with the addition of the excitation system, the next state equation (2.53) could be written in the following form:

$$s\Delta\psi_{fd} = a_{31}\Delta w_r + a_{32}\Delta\delta + a_{33}\Delta\psi_{fd} + a_{34}\Delta v_1$$

......(2.53)

where

$a_{31} = 0$ (not directly affected by the term $\Delta\omega_r$)

$$a_{32} = \frac{\omega_0 R_{fd}}{L_{fd}} m_1 L'_{ads}$$

$$a_{33} = -\frac{\omega_0 R_{fd}}{L_{fd}}\left[1 - \frac{L'_{ads}}{L_{fd}} + m_2 L'_{ads}\right] \qquad ......(2.54)$$

$$a_{34} = -\frac{\omega_0 R_{fd}}{L_{adu}} K_A$$

Since a first-order model for the exciter is used, the overall system order is increased by one, and the new added state is Δv_1. From equation (2.51), we obtain:

$$s\Delta v_1 = a_{41}\Delta\omega_r + a_{42}\Delta\delta + a_{43}\Delta\psi_{fd} + a_{44}\Delta v_1 ......(2.55)$$

where

$a_{41} = 0$(same as a_{31} above), $a_{42} = \dfrac{K_5}{T_R}$

$a_{43} = \dfrac{K_6}{T_R}$, $a_{44} = -\dfrac{1}{T_R}$ $\qquad ......(2.56)$

The accumulated information from preceding equations is sufficient to develop the following complete state-space model for the power system including the excitation system of Fig. (2.10):

$$
\begin{bmatrix} \Delta\dot{\omega}_r \\ \Delta\dot{\delta} \\ \Delta\dot{\psi}_{fd} \\ \Delta\dot{v}_1 \end{bmatrix} = \begin{bmatrix} a_{11} & a_{12} & a_{13} & 0 \\ a_{21} & 0 & 0 & 0 \\ 0 & a_{32} & a_{33} & a_{34} \\ 0 & a_{42} & a_{43} & a_{44} \end{bmatrix} \begin{bmatrix} \Delta\omega_r \\ \Delta\delta \\ \Delta\psi_{fd} \\ \Delta v_1 \end{bmatrix} + \begin{bmatrix} b_1 \\ 0 \\ 0 \\ 0 \end{bmatrix} \Delta T_m
$$

......(2.57)

where

$$
a_{11} = -\frac{K_D}{2H} \quad a_{12} = -\frac{K_1}{2H} \quad a_{13} = -\frac{K_2}{2H}
$$

$$
a_{21} = \omega_0 = 2\pi f_0
$$

$$
b_1 = \frac{1}{2H} \qquad\qquad\qquad ......(2.58)
$$

For coefficients a_{32}, a_{33} and a_{34} , expressions are the same as those in (2.54). and for coefficients a_{42}, a_{43} and a_{44} , they are the same as mentioned in (2.56). with constant mechanical torque input, $\Delta T_m = 0$.

From the completed state-space model of the system, a block diagram is shown in Fig. (2-10). In this model, the characteristics of the system are expressed using the so-called K-constants [3]. The expressions for K_1, K_2 are derived in (2.44) and (2.45) and for K_5, K_6 in (2.49) and (2.50), other constants in the diagram could be found by using the simplified field circuit dynamic equation:

$$
s\Delta\psi_{fd} = a_{32}\Delta\delta + a_{33}\Delta\psi_{fd} + B\Delta E_{fd} \qquad ......(2.59)
$$

By grouping and rearranging terms involving $\Delta\psi_{fd}$, we get

$$\Delta\psi_{fd} = \frac{K_3}{1+sT_F}\left[\Delta E_{fd} - K_4\Delta\delta\right] \qquad \text{......(2.60)}$$

where

$$K_3 = \frac{B}{a_{33}}$$

$$K_4 = \frac{a_{32}}{B} \qquad \text{given that } B = \frac{\omega_0 R_{fd}}{L_{adu}}$$
......(2.61)

$$T_F = -\frac{1}{a_{33}}$$

2.2 Stability Analysis of the Modeled Power System

In this section, the effects of various torque components on the air-gap torque, labeled as ΔT_e which have a direct effect on the rotor speed will be studied.

2.2.1 Effect of field flux linkage variation on system stability

We see from Fig. (2-10) that with constant field voltage ($\Delta E_{fd} = 0$) the field flux variation is caused only by feedback of $\Delta\delta$ through the coefficient K_4. This represents the demagnetizing effect of the armature reaction, (the state variables are shown on the diagram with numbering in brackets).

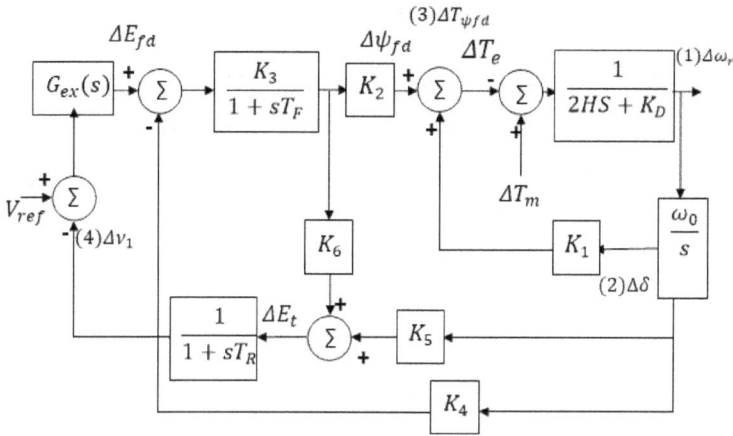

Fig. (2-10) System block diagram with field circuit, exciter and AVR

The demagnetizing effect is represented by the loop through K_4, K_3 and K_2 in Fig. (2-10). It will be defined as ΔT_{ar} which represents the torque component resulting from armature reaction or $\Delta T_{\psi fd}$ where:

$$\frac{\Delta T_{ar}}{\Delta \delta} = \frac{-K_2 K_3 K_4}{1+sT_F} = \frac{K_2 K_3 K_4 \angle -180°}{1+sT_F} \qquad \ldots\ldots(2.62)$$

From the last transfer function we can identify the phase of the electrical torque contribution relative to $\Delta \delta$, which is:

$$\phi_{ar} = -180 - tan^{-1} T_F \omega_{osc} \qquad \ldots\ldots(2.63)$$

where ω_{osc} is the frequency corresponding to the weakly damped electromechanical modes of oscillations mentioned earlier (from 0.5 Hz to about 2 Hz). To demonstrate this effect on the resultant torque at the air-gap, it is better to go back to the torques vector diagram of Fig. (2-6), where the resultant torque component is divided into synchronizing component and damping component. In equ. (2.63), the effect of ΔT_{ar} adds a vector with an angle equals ϕ_{ar}. It is clear that ϕ_{ar} equals $-180°$ plus additional negative angle from the term $(tan^{-1}T_F\omega_{osc})$. Since $(T_F\omega_{osc})$ is positive, this additional angle must be between $0°$ and $90°$. This is shown in Fig. (2-11).

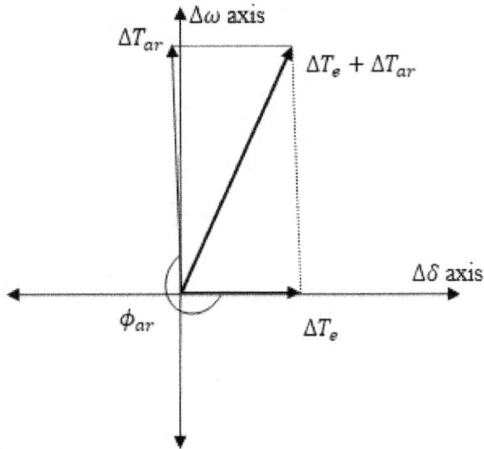

Fig. (2-11) effect of armature reaction on the air-gap torque vector

It is clear from Fig. (2-11) that the effect of armature reaction on the composite torque is to increase damping torque (which is in phase with $\Delta\omega$) and to decrease synchronizing torque (in phase with $\Delta\delta$).

Further analysis could be given by considering the linearized modeled system of Fig.(2-10) in a time-domain analysis where the system is working at operation conditions (given in app.[A]) assuming manual excitation control (constant E_{fd}) and transmission line 2 is out of service. The goal of this analysis is to compute the elements of the system matrix A and input matrix B using the equations (2.54), (2.56) and (2.57) that could be employed when performing time-domain analysis. One can see that when the system is working on constant E_{fd} , the system matrix A is of third order, with states defined as $\Delta\omega$, $\Delta\delta$ and $\Delta\psi_{fd}$. Using the data and conditions given in appendix [A], the calculated elements of the A and B matrix are:

$$A = \begin{bmatrix} 0 & -0.1089 & -0.1250 \\ 376.991 & 0 & 0 \\ 0 & -0.1967 & -0.4133 \end{bmatrix}, B = \begin{bmatrix} 0.1429 \\ 0 \\ 0 \end{bmatrix}$$

Now, when subjecting the system to a step change in its input, the time domain response of $\Delta\omega$ is obtained by the aid of MATLAB simulation environment, and this is shown in Fig. (2-12) below.

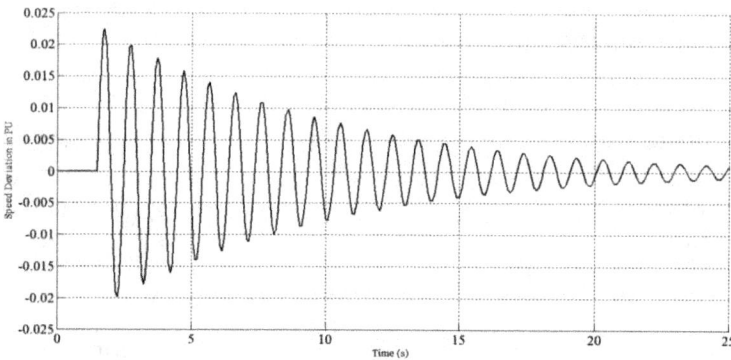

Fig. (2-12) System speed deviation $\Delta\omega$ response to a step change

Eigenvalues are calculated at this operating condition using the standard MATLAB routine, and their values are:

$$\lambda_{1,2} = -0.128 \mp j6.4031$$

$$\lambda_3 \;\; = -0.1877$$

The dominant eigenvalues $(\lambda_{1,2})$ show that the system is stable with damping ratio of ($\zeta = 0.019$) and damped oscillation frequency of ($\omega_d = 1.019 \, Hz$). We can compute K_S and K_D due to armature reaction effect using the obtained values of ζ and ω_d, so:

$$K_S(\Delta\psi_{fd}) = -0.00172 \;\; \text{per unit torque/rad}$$

$$K_D(\Delta\psi_{fd}) = \;\; 1.53 \quad \text{p.u. torque/p.u. speed change}$$

The effect of field flux variation (i.e., armature reaction) is thus to reduce the synchronizing torque slightly and to add a damping torque component.

2.2.2 Effect of excitation system & AVR on synchronizing and damping torque components

With the addition of excitation system and Automatic Voltage Regulator (AVR), a new torque component is added to the air-gap resultant torque. This component results from the K_5 and K_6 loops shown in Fig. (2-10). The field flux variations are caused by the field voltage variations, in addition to the armature reaction [1].

From block diagram of Fig.(2-10), the torque component due to the field flux variation (denoted as ΔT_{e1}) is defined as follows :

$$\Delta T_{e1} = \frac{K_3 K_2}{1 + sT_F}\left[-K_4\Delta\delta - \frac{G_{ex}(s)}{1 + sT_R}(K_5\Delta\delta + K_6\Delta\psi_{fd})\right]$$

......(2.64)

where

$$\Delta T_{e1} = \Delta T_e + \Delta T_{ar} + \Delta T_{exc}$$

By grouping the terms and rearranging

$$\Delta T_{e1} = \frac{-K_2 K_3 [K_4 (1+sT_R) + K_5 G_{ex}(s)]}{s^2 T_F T_R + s(T_F + T_R) + 1 + K_3 K_6 G_{ex}(s)} \Delta\delta \quad \ldots\ldots(2.65)$$

The effect of the excitation system on damping and synchronizing torque component is primarily influenced by K_5 and $G_{ex}(s)$. Now by taking the same ratings and operating condition given in (app. [A]), and for a thyristor exciter:

$$G_{ex}(s) = K_A \qquad\qquad \ldots\ldots(2.66)$$

From equ. (2.65), assuming that the rotor oscillation frequency is 10 rad/s (1.6 Hz), with $s = j\omega = j10$, $K_5 = -0.12$ and $K_A = 200$:

$$\Delta T_{e1} = \frac{11.1 - j0.8}{17.18 + j19.3} \Delta\delta = 0.2804\Delta\delta - 0.3255(j\Delta\delta)$$

knowing that $\Delta\omega_r = s\Delta\delta/\omega_0 = j\omega\Delta\delta/\omega_0$ so with $\omega = j10$ rad/s, the damping torque coefficient is = -12.27 pu torque/pu speed.

The first term is associated with synchronizing torque, while the second one is associated with damping torque. Thus the effect of the AVR is to increase the synchronizing torque component and decrease the damping torque component when K_5 is negative (which is the common case of operation).This effect could be verified through constructing the torque vector diagram with the addition of the torque component developed by the excitation system. For a typical power system oscillation frequencies ranging from 0.2 Hz to 2 Hz, the range of phase of this component is between -20 to -90 degrees with respect to $\Delta\delta$, and it tends to be closer to -90° for modern, high gain (large K_A), fast response static excitation systems. So the diagram will become as in Fig. (2-13). It is clear from Fig. (2-13) that the effect of the exciter is to add a positive synchronizing torque component, but it adds a negative damping torque component, and at high external system reactance and high generator output, this case is commonly met, so this leads to a sustained or increasing oscillation amplitude which means that the system becomes critically stable or unstable

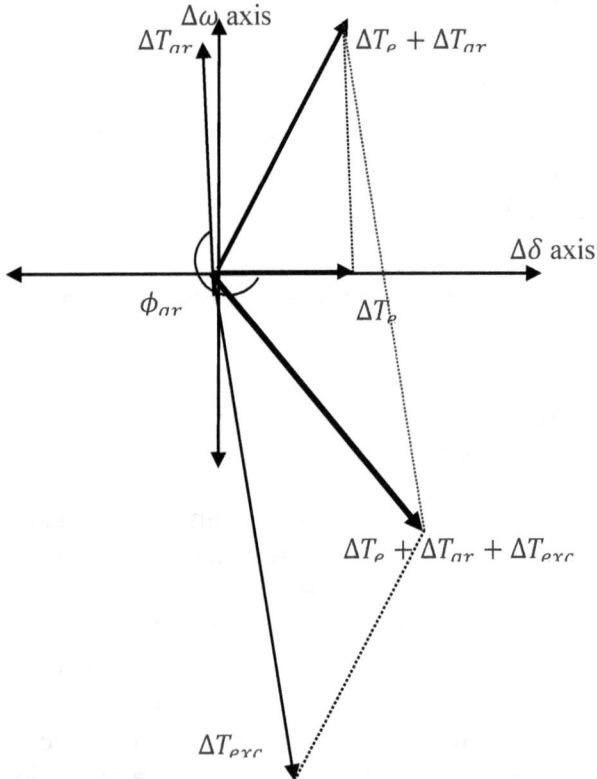

Fig. (2-13) Torque vector diagram with effects of excitation system & AVR

The system matrices including the exciter and AVR calculated using equation (2.57) are:

$$A = \begin{bmatrix} 0 & -0.1089 & -0.1250 & 0 \\ 376.991 & 0 & 0 & 0 \\ 0 & -0.1967 & -0.4133 & -27.4175 \\ 0 & -7.2968 & 20.9090 & -50 \end{bmatrix}, B = \begin{bmatrix} 0.1429 \\ 0 \\ 0 \\ 0 \end{bmatrix}$$

The state of instability which is obvious from a time-domain simulation of the linearized state space model, as shown in Fig. (2-14) under the same conditions in case of Fig. (2-11) but with the addition of excitation system and AVR.

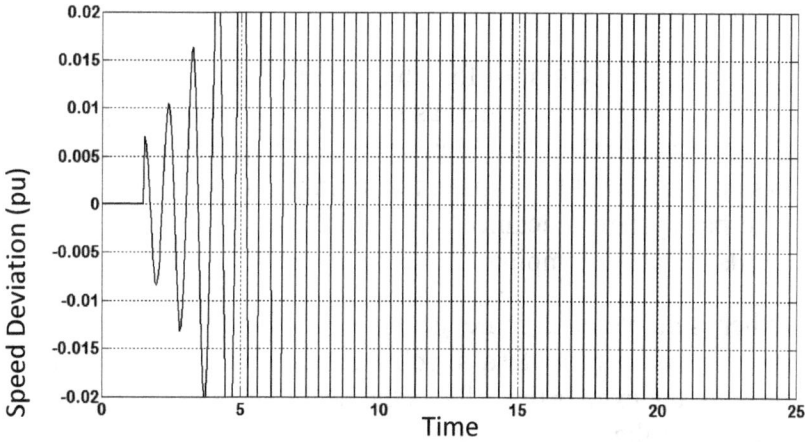

Fig. (2-14) Speed deviation $\Delta\omega$ response due to a step change for a complete state-space model of SMIB system

Now the eigenvalues are:

$\lambda_1, \lambda_2 = 0.5061 \pm j7.2327$

$\lambda_3 \quad = -30.8755$

$\lambda_4 \quad = -20.5500$

So the system is unstable through an oscillatory mode of 1.15 Hz and a damping ratio of -0.0698. The main conclusion from the previous two subsections is that an AVR, which reacts only to the voltage error, weakens the damping introduced by the machine inertial forces and field windings. In the extreme case of a heavily loaded generator operating on a long transmission link, a large increase in the voltage controller gain may result in net negative damping leading to an oscillatory loss of stability [5].

One possible solution for this problem is to make K_A (exciter gain value) as high as possible without causing un-damped oscillation. But high-gain, fast response excitation systems are good for transient damping. But with very high external system reactance, even this restriction of K_A still yields negative damping torque component [1].

Other suitable and more accepted method to solve this problem is to provide a supplementary torque component through the exciter that offsets the negative damping torque caused by the excitation system. This device that controls this damping process is referred to as the *power system stabilizer*.

2.3 Power System Stabilizer

The main idea of power system stabilization is to recognize that in the steady state period, when the speed deviation $\Delta\omega$ is zero or nearly zero, the voltage controller should be driven by the voltage error ΔV_t from the AVR only. However, in the transient state the generator speed is not constant, the rotor swings, and ΔV_t undergoes oscillations caused by the change in rotor angle. The task of the PSS is to add an additional signal which compensates the ΔV oscillations and provides a damping component that is in phase with $\Delta\omega$. In other words, the PSS transfer function $G_{PSS}(s)$ should have appropriate phase lead between the exciter input and the electrical torque. Since the purpose of a PSS is to introduce a damping torque component, a suitable signal to use as an input for the PSS is the speed deviation $\Delta\omega$ [1].

Power system stabilizers are designed with various types of inputs: They are speed, frequency, power, accelerating power and integral of accelerating power. The PSS may acquire these quantities from generator terminal voltage and current measurements. Currently, new methods are based on deriving a synthetic speed measurement (integral of accelerating power) from generator terminal voltage and current measurements [26].

The general form of a PSS consists of three main parts:

1. Gain block.
2. Wash-out block.
3. Phase compensation block.

It also has a block diagram as shown in Fig. (2-15).

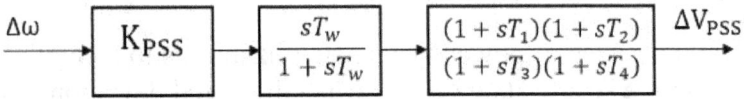

$\Delta\omega \longrightarrow \boxed{K_{PSS}} \longrightarrow \boxed{\dfrac{sT_w}{1+sT_w}} \longrightarrow \boxed{\dfrac{(1+sT_1)(1+sT_2)}{(1+sT_3)(1+sT_4)}} \xrightarrow{\Delta V_{PSS}}$

Fig. (2-15) general configuration of power system stabilizer PSS

The first block represents the stabilizer gain factor which determines the amount of damping introduced by the PSS. Ideally, the gain should be set at a value corresponding to maximum possible damping.

The *washout* block serves as a high-pass filter, with time constant high enough to allow signals associated with oscillations in ω_r to pass unchanged. Without it, steady changes in speed would modify the terminal voltage of the machine.

The *phase compensation* block provides the appropriate phase characteristics to compensate the phase between the exciter input and the generator electrical (air-gap) torque. The transfer function of the given stabilizer is:

$$G_{PSS} = \frac{\Delta v_{PSS}}{\Delta \omega_r} = K_{PSS}\frac{sT_w}{1+sT_w}\frac{(1+sT_1)(1+sT_2)}{(1+sT_3)(1+sT_4)} \qquad \ldots\ldots(2.67)$$

The stabilizer gain K_{PSS} is responsible mainly for the amount of damping introduced to the system, which has a large impact on the speed of response of the system, so it is kept fixed in the search process in this study. The stabilizer frequency characteristic is adjusted by varying the time constants T_1, T_2, T_3 and T_4 . The value of T_w is not critical and maybe in the range of 1 to 20 seconds. A power system stabilizer can be most effectively applied if it is tuned with an understanding of the associated power system characteristics (i.e. field flux variation, system parameters, etc.) and the function to be performed by the stabilizer. A knowledge of the modes of power system oscillation to which the stabilizer is to provide damping would define the range of frequencies over which the stabilizer must operate. Simple analytical models, such as that of a single machine connected to an infinite bus, can be useful in determining the frequencies of local mode oscillations during the planning stage of a new plant [10].

2.4 Traditional Method of Power System Stabilizer Design

Power system stabilizer typically works on phase compensation and adjustment and this is the main task in PSS tuning. If the exciter transfer function $G_{ex}(s)$ and the generator transfer function between ΔE_{fd} and ΔT_e was pure gain, a direct feedback $\Delta\omega_r$ would result in a damping torque component. However, in practice both generator and the exciter (depending on its type) exhibit frequency dependent gain and phase characteristics. Therefore, the PSS transfer function, $G_{PSS}(s)$ should have appropriate phase compensation circuit to compensate the phase lag through the generator excitation system and power system, such that PSS provides torque changes in phase with speed changes, this phase lag is evaluated from the linearized model of the SMIB system through the path from Δv_1 to ΔT_e which could be considered as the path of the supplementary torque component introduced from the PSS. So conventional approach to PSS tuning is to make it an exact inverse of the exciter and generator phase characteristics to be compensated [1], [4]. Fig. (2-16) shows the system block diagram which is an extension of that shown in Fig. (2-10) with the PSS.

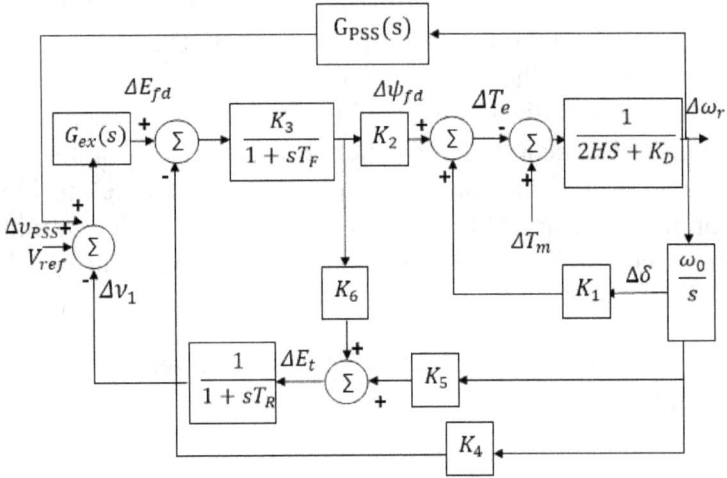

Fig. (2-16) System block diagram with AVR and PSS

To illustrate the design process of power system stabilizer, the previous system will be considered, with the same parameters and operating conditions. Now since the given T_R value is very small compared to T_F, we will neglect its effect in examining the PSS performance. This simplifies the analysis without loss of accuracy. Now from block diagram of fig.(2-16), $\Delta\psi_{fd}$ due to PSS is given by :

$$\Delta\psi_{fd} = \frac{K_3 K_A}{1+sT_F}\left(-K_6\Delta\psi_{fd} + \Delta v_{PSS}\right) \qquad ......(2.68)$$

$$\frac{\Delta\psi_{fd}}{\Delta v_{PSS}} = \frac{K_3 K_A}{sT_F + 1 + K_3 K_6 K_A} \qquad ......(2.69)$$

Using the values of the considered parameters:

$$\frac{\Delta\psi_{fd}}{\Delta v_{PSS}} = \frac{66.34}{2.42s + 28.74}$$

Now the PSS phase compensation required to produce damping at a given rotor oscillation frequency of 10 rad/s can be recognized as follows: for $s = j\omega = j10$,

$$\frac{\Delta\psi_{fd}}{\Delta v_{PSS}} = \frac{66.34}{28.74 + j24.2}$$

for the block diagram of Fig.(2-16),the torque effect of the PSS is:

$\Delta T_{PSS} = \Delta T_e$(due to PSS) $= K_2 * \Delta\psi_{fd}$(due to PSS), and at ω_{osc}=10 rad/s,

$$\frac{\Delta T_{PSS}}{\Delta v_{PSS}} = K_2\left[\frac{66.34}{28.74 + j24.2}\right] = \frac{58.03}{28.47 + j24.2}$$

$$= 1.553\angle - 40.3651°$$

If ΔT_{PSS} has to be in phase with $\Delta\omega_r$ (i.e., pure damping torque), the $\Delta\omega_r$ signal should be processed through a *phase-lead network* so that the signal is advanced by $\theta = 40.36°$ at a frequency of oscillation of 10 rad/s. The amount of damping introduced by PSS depends on its transfer function at that frequency.

2.5 Effect of PSS on Damping & Synchronizing Torque

The principle of the power system stabilizer is to cancel the phase lag introduced by the excitation system with the right amount of lead compensation so the torque applied on the shaft by the excitation control effect will be in phase with speed deviation. This theory could be illustrated by expanding Fig. (2-12) to include the torque component developed by the PSS, This is illustrated in fig. (2-17) below.

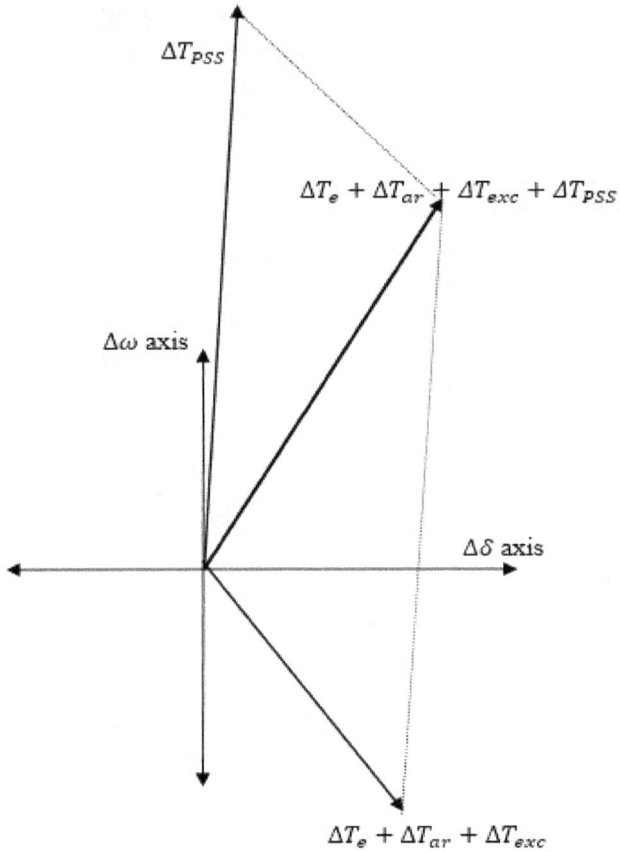

Fig. (2-17) Torque vector diagram with effect of PSS

It is evident from Fig. (2-17) that the resultant torque is shifted back to the positive damping region, which means that the damping torque component is in phase with the speed deviations ($\Delta\omega_r$). It produces positive damping torque component and eventually, suppressing the low frequency oscillation and preventing the system from reaching the instability state condition.

2.6 System State Matrix

The linear dynamic model of the power system without PSS was discussed in detail in Section 2.4. Following the same convention, the representation of the above composite system with the inclusion of the PSS can be described by:

$$\dot{x} = Ax(t) + Bu(t) \qquad \qquad ......(2.70)$$

$x(t)$ is the state vector, $\mathbf{A}$ and $\mathbf{B}$ are the state and input matrices, respectively and $u(t)$ is the input vector. The above equation defines the system that includes the PSS model given below in Fig. (2-18)

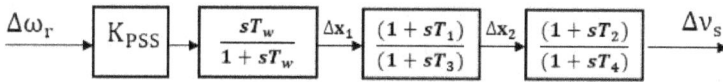

Fig. (2-18) Power system stabilizer block diagram

Since this modeled PSS is of third order, when added to the linearized power system model, it will raise the order of the system by three. Now from block diagram of Fig.(2-18):

$$\Delta x_1 = \Delta \omega_r \left[\frac{sK_{PSS}T_W}{1+sT_W} \right] \qquad ......(2.71)$$

$$\Delta x_2 = \Delta X_1 \left[\frac{(1+sT_1)}{(1+sT_3)} \right] \qquad ......(2.72)$$

$$\Delta v_s = \Delta X_2 \left[\frac{(1+sT_2)}{(1+sT_4)} \right] \qquad ......(2.73)$$

Now, by substituting the term $\Delta\omega_r$ from (2.57), regrouping and rearranging the other terms to make these three newly added states $(\Delta x_1, \Delta x_2$ and $\Delta v_s)$ together with the other state variables and system parameters, the overall state-space model of the system with the PSS will become of 7th order, the new state vector is :

$$X(t) = \begin{bmatrix} \Delta\omega_r \Delta\delta & \Delta\psi_{fd}\Delta v_1\Delta X_1\Delta X_2\Delta v_s \end{bmatrix} \quad \ldots\ldots(2.74)$$

The system matrices A and B of the system given by equation (2.70) are:

$$\dot{x} = \begin{bmatrix} 0 & a_{12} & a_{13} & 0 & 0 & 0 & 0 \\ a_{21} & 0 & 0 & 0 & 0 & 0 & 0 \\ 0 & a_{32} & a_{33} & a_{34} & 0 & 0 & a_{37} \\ 0 & a_{42} & a_{43} & a_{44} & 0 & 0 & 0 \\ 0 & a_{52} & a_{53} & 0 & a_{55} & 0 & 0 \\ 0 & a_{62} & a_{63} & 0 & a_{65} & a_{66} & 0 \\ 0 & a_{72} & a_{73} & 0 & a_{75} & a_{76} & a_{77} \end{bmatrix} x(t)$$

$$+ \begin{bmatrix} b_{11} \\ 0 \\ 0 \\ 0 \\ b_{51} \\ b_{61} \\ b_{71} \end{bmatrix} u(t)$$

Given that $u = [\Delta T_M]$

Where matrices' coefficients are specified below:

$$a_{12} = -\frac{K_1}{2H}, \quad a_{13} = -\frac{K_2}{2H}, a_{21} = 2\pi f_0$$

$$a_{32} = \frac{\omega_0 R_{fd}}{L_{fd}} m_1 L'_{ads} ,$$

$$a_{33} = -\frac{\omega_0 R_{fd}}{L_{fd}} \left[1 - \frac{L'_{ads}}{L_{fd}} + m_2 L'_{ads} \right]$$

$$a_{34} = -\frac{\omega_0 R_{fd}}{L_{adu}} K_A \quad , a_{37} = K_A \frac{\omega_0 R_{fd}}{L_{adu}},$$

$$a_{42} = \frac{K_5}{T_R} \quad , \quad a_{43} = \frac{K_6}{T_R} \quad , a_{44} = -\frac{1}{T_R}$$

$$a_{52} = K_{PSS} a_{12} , \quad a_{53} = K_{PSS} a_{13} , \quad a_{55} = -\frac{1}{T_w}$$

$$a_{61} = a_{51} \frac{T_1}{T_3} \quad , \quad a_{62} = a_{52} \frac{T_1}{T_3} \quad , \quad a_{63} = a_{53} \frac{T_1}{T_3}$$

$$a_{65} = a_{55} \frac{T_1}{T_3} + \frac{1}{T_3} , a_{66} = -\frac{1}{T_3}$$

$$a_{71} = a_{61} \frac{T_2}{T_4} \quad , \quad a_{72} = a_{62} \frac{T_2}{T_4} \quad , \quad a_{73} = a_{63} \frac{T_2}{T_4}$$

$$a_{75} = a_{65} \frac{T_2}{T_4} \quad , \quad a_{76} = a_{66} \frac{T_2}{T_4} + \frac{1}{T_4} , a_{77} = -\frac{1}{T_4}$$

$$b_{11} = \frac{1}{2H} \quad , \quad b_{51} = b_{61} = b_{71} = \frac{K_{PSS}}{2H}$$

2.7 Traditional Design Results

Using the traditional PSS design method explained in section (2.4), the PSS time constants obtained were as in table (2-1), given that the PSS gain block value is (**9.5**) and washout time constant T_w is (**10**) sec.

Table (2-2) gives the performance index (IAE) value for the speed deviation error (e_i), when subjected to a step change and calculated using equ. (2.75) and the amount of phase of the PSS transfer function (from the speed deviation input signal to the PSS voltage output signal) when the machine is supplying nominal loading of $P_e = 0.8$ per unit and $Q_e = 0.3$ per unit.

Performance index $= IAE = \frac{1}{N}\sum_1^N |e_i|$ ......(2.75)

where $e_i = \Delta\omega_r$

Table (2-1) Traditional PSS design method results

Design method	T1	T2	T3	T4
Classic PSS	0.1540	0.1540	0.330	0.1330

Table (2-2) Performance index and phase of the traditional PSS design

Design method	Performance Index ($\times 10^{-2}$)	phase (in deg. Leading)
Classic PSS	0.9084	42.6805

2.8 Linearized System Simulations

The complete simulation block diagram of the linearized SMIB model is given in Fig. (2-19), system data are given in appendix[A]. Different configurations are assembled based on the removal and addition of these components to the power system model.

Fig. (2-19) Power system model in Matlab®/Simulink®

First case: the system is operating at manual excitation (without using an AVR unit). Linear simulation of the system was performed. Fig.(2-20) shows the system step response for a period of 0.05 sec. operating at nominal loading and without AVR at the speed deviation (Δω) output.

Fig. (2-20) System without AVR speed deviation response

The simulation also was performed to find the rotor angle deviation in p.u., Fig. (2-21) shows delta deviation response

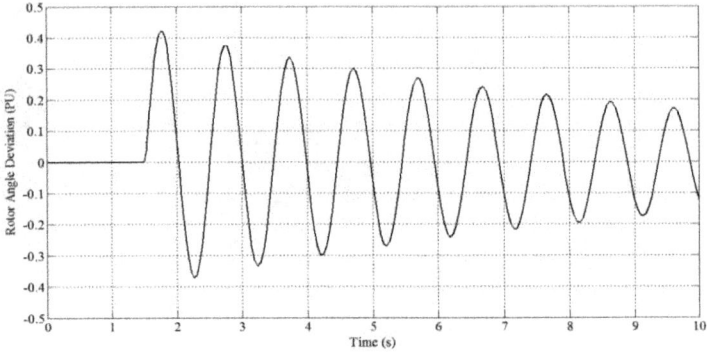

Fig. (2-21) Rotor angle deviation response for system without AVR

Second case: the system is fitted with an AVR unit. Linear simulation of the system was performed to find the open-loop (without a PSS compensator) response. Fig (2-22) shows the speed deviation response when subjecting the system to the same disturbance as in case (1).

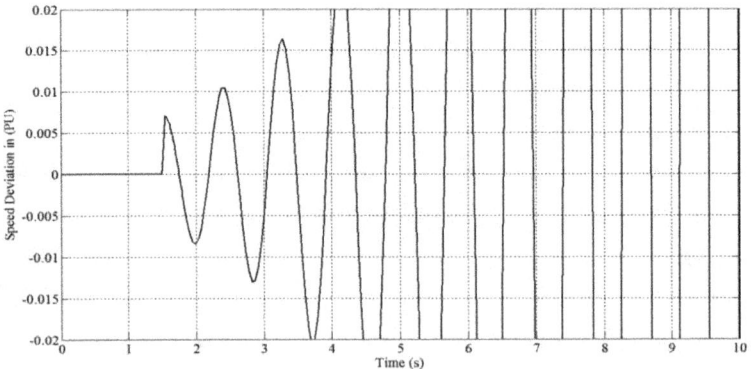

Fig. (2-22) System with an AVR speed deviation response

The response of the rotor angle deviation ($\Delta\delta$) was also taken in the simulation and it is shown in Fig. (2-23).

Fig. (2-23) Rotor deviation response for system with an AVR

It is clear from the step response on both the speed and delta deviation that the system is instable at the operating condition and after the addition of an AVR unit when subjected to external disturbance.

Third case: (system with PSS) The parameters of the PSS were determined based on the classic tuning approach. Linear simulations were performed on the system with the PSS and under the same operating conditions as in case 1 & 2. Fig. (2-24) shows the speed deviation response for the system with the PSS.

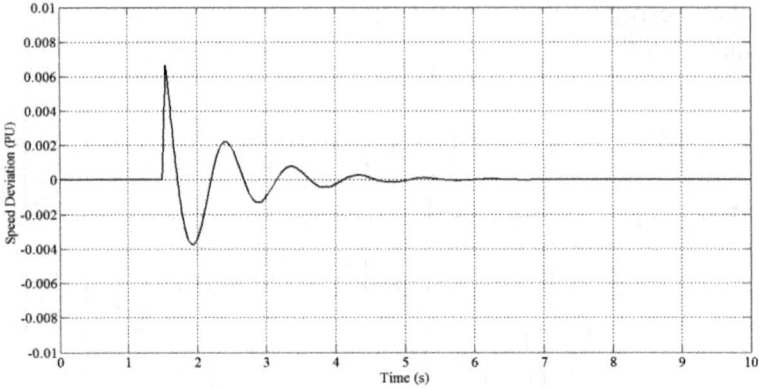

Fig. (2-24) System with Traditional - PSS speed deviation response

In Fig. (2-24), the rotor angle response was taken for the same system in case 3 with the same applied disturbance. It is evident from both Figs. (2-24 and 2-25) in addition to the eigen values analysis that the system restored its stability to the external disturbance at the operating conditions with the application of the PSS.

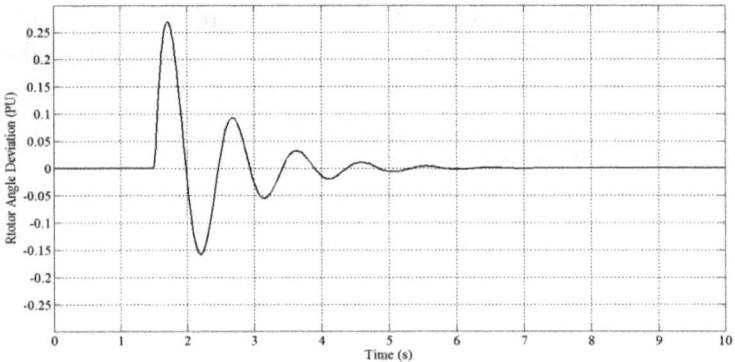

Fig. (2-25) Rotor angle deviation response for system with Classic PSS

CHAPTER THREE

POWER SYSTEM STABILIZER DESIGN BASED ON GENETIC ALGORITHM

In this chapter, a genetic-algorithm is used to design power system stabilizer (GA-PSS). The design process is based on tuning the parameters of the PSS to meet specific objective function. A new damping optimization method within the genetic algorithm is developed to ensure the electro-mechanical modes of oscillation damping in the system.

3.1 Genetic Algorithms Based Optimization

Optimization is a general technique used in solving various problems of engineering and sciences .Genetic algorithms (GAs) are a subclass of evolutionary optimization algorithms. They provide a powerful tool for optimization problems by imitating the mechanisms of natural selection and genetics which operate on a population of potential solutions (applying the principle of *Survival of the Best* to produce better and better guesses for the solution) in each generation, a new set of approximations is created by selecting the individuals according to their level of fitness in the problem domain and breeding them together using operators borrowed from natural genes [4], [8].

3.2 Biological Evolution Process

In nature, all living organisms basically consist of cells. Every cell consists of a set of chromosomes. A chromosome is a long, composite string of DNA and serves as a model for the whole organism.

Reproduction is the process during which new chromosomes are created. The first thing that occurs in the creation of new chromosomes is called *crossover* or *recombination*. During crossover genes from the parent chromosomes recombine and create new chromosomes. The other important operator that takes place during reproduction is called *mutation*. During mutation, basically, a small change is incorporated in the elements of DNA.

Survival is the measure of fitness of an organism. The individual that has better survival traits will survive for a longer period of time. This in turn provides it a better chance to produce offspring with its genetic material; therefore, after a long period of time, the entire population will consist of lots of genes from the superior individuals and less from the inferior individuals. In a sense, the fittest survived and the unfit died out. This is called natural selection [30], [31].

The basic structure of a natural DNA is shown in Fig. (3-1).

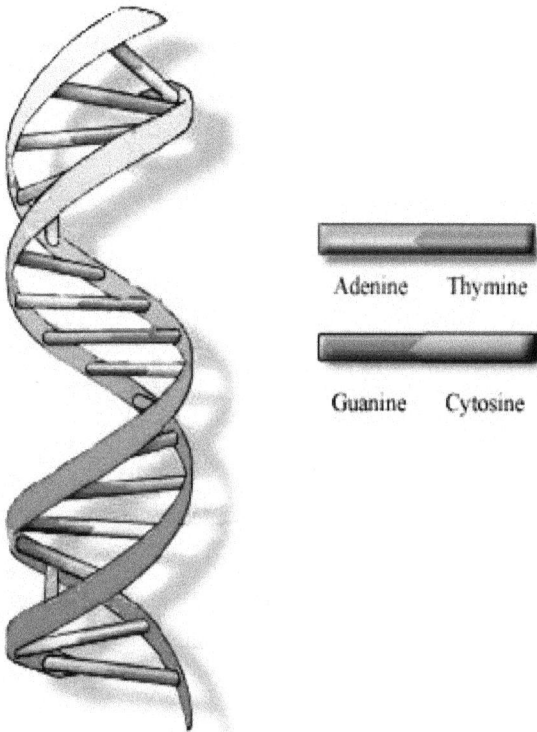

Fig. (3-1) Structure of a DNA

3.3 Components of a Genetic algorithm (GA)

The components of a genetic algorithm along with the main cycle of the genetic search process are shown in Fig. (3-2). Details on these components and operators are given next,

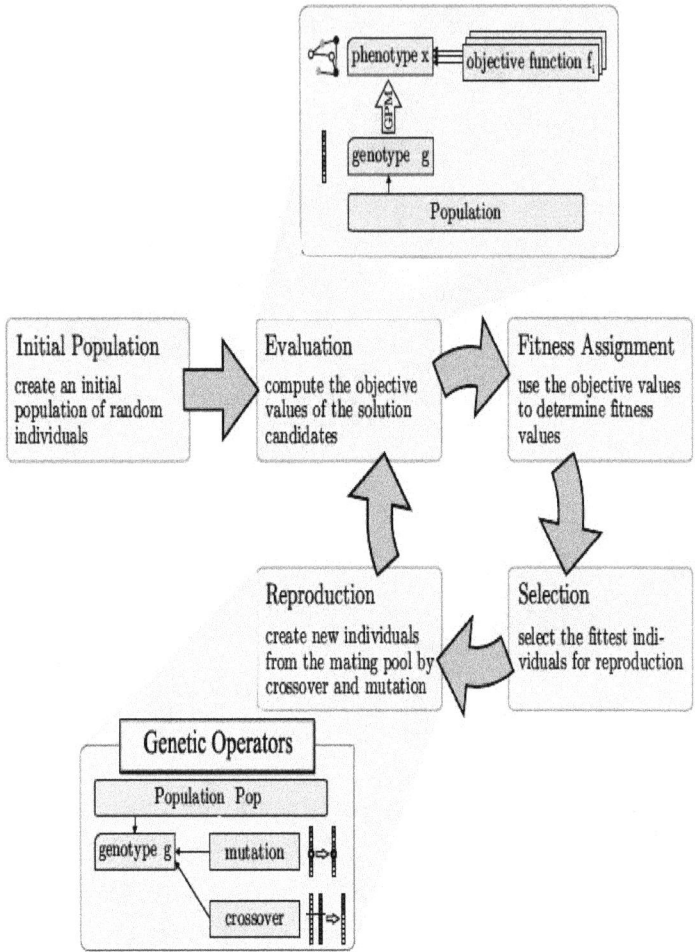

Fig. (3-2) Basic cycle of a genetic algorithm

3.3.1 Population: One of the most important factors which control the genetic algorithm operation are the parameters of the population, namely population size (total number of individuals) and maximum number of generations to be produced. Larger population sizes require more computation time but they can converge more quickly. For instance, for a larger population size there will be more guesses of the position of the global maximum is for the current population and hence it is more likely that there will be at least one good guess.

3.3.2 Encoding method: This is the method which must be used to represent an individual in the population, and links the real-world phenotype (original problem context) to the GA- domain genotype (problem-solving space where evolution takes place). This encoding/decoding function must be also invertible. Choosing a certain type of encoding method depends mainly on the problem to be optimized. Mostly, in a genetic algorithm, a chromosome corresponds to a vector of real parameters. There are several types of encoding methods which may be used to represent the individuals in a population [8].

a. Binary encoding: Each chromosome is expressed as a string of 1's and 0's. Binary encoding is the most common because the beginning of GA studies was with this type of encoding.

Chromosome A :	101100101100101011100101
Chromosome B :	111111100000110000011111

b. Permutation encoding: The chromosome is represented as a string of numbers (vector of integers), which represents number in a sequence. Permutation encoding can be used in ordering problems, such as travelling salesman problem or task ordering problem.

Chromosome A :	1 5 3 2 6 4 7 9 8
Chromosome B :	8 7 5 9 3 0 4 6 0

c. Real Values Encoding: This method represents each chromosome with a string of some values which can be real floating point numbers, letters or even characters, mostly encoding is linked directly to the problem itself. This method has the advantage of requiring less storage than the binary encoding because a single floating-point number represents the variable instead of n_{bits} integers. The continuous GA is inherently faster than the binary GA, because the chromosomes do not have to be decoded prior to the evaluation of the cost function.

Chromosome A :	2.324 5.534 1.665 0.405
Chromosome B :	0.002 4.432 4.532 1.432

The strength of real-coded GA includes increased efficiency (where bit strings do not need to be converted to real numbers for every function evaluation), increased precision (since a real-number representation is used, there is no loss of precision due to the binary representation) and greater freedom to use different mutation and crossover techniques based on the real representation [32], [33].

3.3.3 Selection

Selection process is based on the *Survival of the Fittest* rule, which in nature translates into rejecting the chromosomes with the least fitness. Only the best ones are selected to continue, while the rest will be deleted with each iteration or generation. In the application of genetic algorithm as optimization tool, there are many ways to perform the selection process, but the most common method is the fitness-proportionate selection. In this approach, chromosomes are chosen based on how fit they are (as computed by the fitness function) relative to the other members of the population. Then, selection is used to choose which chromosome should survive to form a "mating pool", more fit individuals end up with increasing number of copies of themselves in the mating pool so that their effect at the formation of the next generation increases each time [34].

Fitness-proportionate selection has several types that include:

a. **Roulette-wheel selection** : this method relies on proportional selection which is based on calculating the selection probability P_{si} as :

$$P_{si} = \frac{f_i}{\Sigma f_i}$$

$$\ldots\ldots (3.3)$$

where f_i is the fitness of the i_{th} chromosome and Σf_i is the summation of the fitnesses related to all of the individuals in the population; respectively.

The result is that the chromosomes with higher fitness among others get higher selection probabilities during the reproduction phase. Since the population size is kept fixed during the GA process, the sum of the probability of each string being selected for the mating pool must be one. Then the associated fitnesses are represented on a wheel in terms of percentages, with each section resembles the selection probability P_{si} of the ith chromosome, as shown in Fig. (3-3).

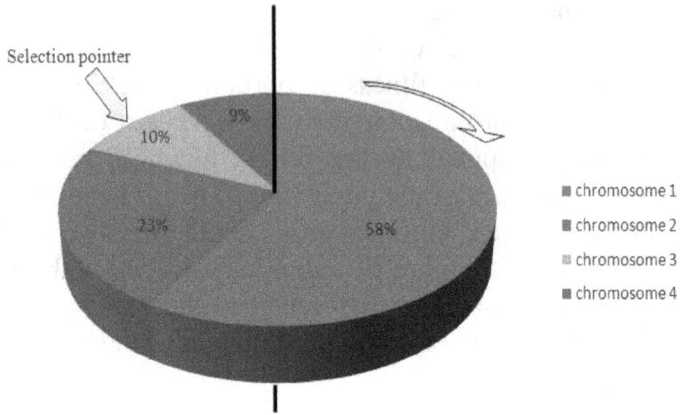

Fig. (3-3) A Roulette-wheel marked for four chromosomes according to their fitness values

For the example of Fig. (3-3), the first chromosome has a higher fitness than the others, so it is expected that the roulette-wheel selection will choose this chromosome or individual more than any other individual. At the same time, this process may choose some relatively unfit individuals and place them in the mating pool due to its probabilistic nature. By repeatedly spinning the roulette wheel, individuals are chosen to fill intermediate population [29], [36].

b. **Stochastic uniform selection:** In this method, a line is laid out in which each parent chromosome is assigned to a section of this line with a length proportional to its scaled value. This algorithm moves along the line in steps of equal size. At each step, the algorithm selects a parent from the corresponding section.

c. **Tournament selection:** Tournament selection chooses each parent by choosing tournament size players randomly and then choosing the best individual (the one with the highest fitness) out of that set to be a parent. This parent is permitted to reproduce. Tournament size must be at least 2. This process is repeated until the mating pool is full. Larger tournaments may also be used, where the best of η randomly chosen individuals is copied into the mating pool.

d. **Rank selection:** Rank selection gives the population a certain rank first, then every chromosome receives a fitness value from this ranking, the worst one will have fitness 1, the second worst 2, etc. and the best will have fitness N where N is the number of the chromosomes in population.

One important feature of selection is to include the best solution (individual) in the next generation without being modified by other genetic operators, This is called *Elitism*, and it copies the best input chromosome (or few chromosomes) to the new population, The rest of the selection is done by using any of the above mentioned ways. This attribute is useful in preventing the loss of the best found solution and can rapidly increase the performance of the genetic algorithm [37].

3.3.4 Reproduction

The formation of what is referred to as "mating pool" is based on the selection process, which defines the individuals that will be used to create the next generation. Afterwards, the reproduction phase starts. The objective of reproduction is to allow information stored in strings with good fitness values to survive into the next generation which is developed from selected pairs of parents' strings and the use of other explorative operators: *Crossover* and *Mutation*. Crossover selects genes from parent chromosomes and produces an offspring. Mutation is performed after the crossover, it randomly changes the new offspring at a certain string point. These two processes will create the new generation [35], [38].

3.3.4.1 Crossover

This is a procedure wherein a selected parent chromosome is broken into segments at certain points and some of these segments are exchanged with corresponding segments of another chromosome. This process has several types to implement in genetic algorithms .These are:

   a. **Single-point Crossover**: In this type, a certain point in the string is selected, and beyond this point all genes in either chromosome are exchanged, which results in the children (offspring) chromosomes Fig. (3-4).

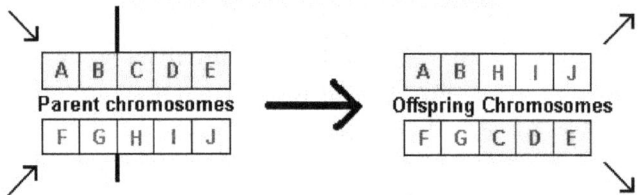

Fig. (3-4) Single-point Crossover (Genetic Mating)

   b. **Double-point Crossover**: This type defines two points on both chromosomes, then all the genes in either chromosome between these points are swapped, giving the new offspring as shown in Fig. (3-5).

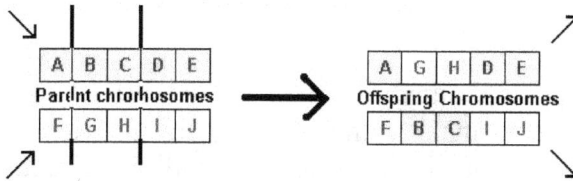

Fig. (3-5) Double-points crossover (Genetic Mating)

c. **Multi-point Crossover**: Here, more than two points are selected to perform crossover at the parents' strings. A number of crossover positions are selected randomly on the string, then the genes between successive crossover points are exchanged between the two parents to produce the new offspring, this is shown in Fig. (3-6).

d.

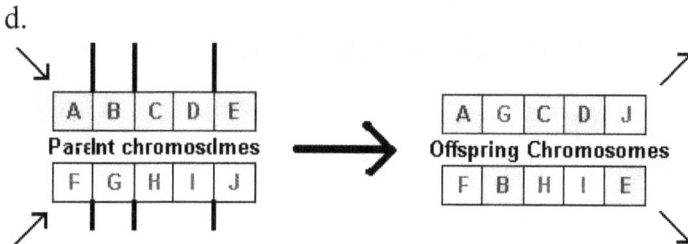

Fig. (3-6) Multiple-points crossover (Genetic Mating)

Usually, crossover is not applied to *all* pairs of chromosomes which are selected for mating. A random choice of pairs is made based on a probability of the application of crossover. This is called *probability of crossover (Pm)* and it is usually in the range of (0.6 to 1.0) [39].

3.3.4.2 Mutation

This is the other vital operator in reproduction process. It is used to make some diversity in the search space and, without it the selection operator will soon choose only the fittest individuals at early stages which will slow down the search and eventually stops at a local extreme solution. This operator deliberately changes a random bit in a chosen chromosome which produces an offspring that is different from the parent. Types of mutation include:

- **Flip-bit (gene)**: This type shown in Fig. (3-7) alters a randomly chosen gene to produce an offspring; This is used only for binary encoded chromosomes.

Fig. (3-7) Mutation operator

- **Boundary**: This type replaces the gene with either a lower or upper bound randomly, and it can be used for integer and float genes.
- **Uniform**: Here, the value of the randomly chosen gene is replaced with a value selected between the upper and lower bounds for that gene.

As in the crossover routine, Pm is the probability of occurrence of mutation, Too low values of Pm will slow down the search, Thus, it may stuck in a local extreme or "premature-convergence" while too high values of *Pm* will prevent the search from finding an optimum solution by continuously breaking the ties in the chromosomes. Typical range of mutation rate is between (0.001 to 0.1) [40], [41].

3.3.5 Stopping criteria:

The search process will continue in a loop. Each new generation is subjected to the same genetic operators depicted in the sections above. By choosing the appropriate boundaries and limits for the factors that are involved in the genetic operators, it is most likely for the search to converge to an optimum solution. Convergence should be monitored so as not to make the search stuck in an infinite loop. In each generation, if there is improvement in the solution, the value of the solution is stored as the current best solution. Stopping criteria is based on the convergence of the solution toward a best possible outcome. This could be performed as a comparison at the end of each iteration. If the population in the generation produces an output that is close enough or equal to the desired answer then the problem of optimization has been solved. Otherwise, the search stops indicating that successive iterations no longer produce better solution. Other termination criterion is to have a maximum number of iterations (generations) and the best solutions stored so as to have a best possible solution and prevent the GA from looping without an end.

3.4 Genetic Algorithm-Based Design of Power System Stabilizer

The modeling of the power system with the addition of PSS has been explained in details in chapter (2). The main problem of the design was to find an optimum settings for the time-constants in the PSS transfer function, namely, $(T_1$ to $T_4)$, which in turn will decide the proper amount of the supplementary damping provided by the PSS through the exciter. In this study, it has been decided to employ a genetic algorithm-based technique in order to find a proper set of values for the time-constants of the PSS. The optimization methodology consists of the following steps in details:

3.4.1 Individuals definition

This step defines the parameters to be optimized in GA and their chromosomal representations. In this study, a modified real-parameter encoding is chosen. The reasons behind this choice are [32]:

a. Bit strings do not need to be converted for every fitness evaluation.
b. Since the parameters are originally real-numbers, the precision is not lost during the conversion.
c. It gives greater flexibility in applying different GA operation techniques.

Basically, this approach gives greater efficiency, precision and freedom to implement on floating numbers. Each individual (chromosome) is represented in base 10 genes. Parameters' set is encoded in real floating numbers. The string is 24 genes long and it represents the four time-constants in floating numbers. The value of each time-constant or trait is bounded between a minimum and a maximum value. These boundaries are kept fixed during the search process. Fig. (3-8) shows an example of a randomly generated chromosome.

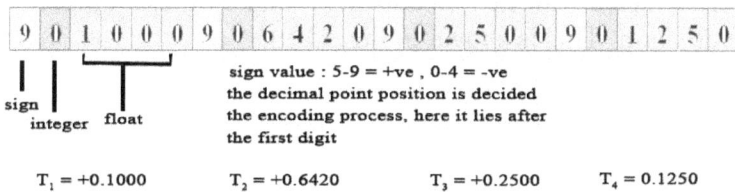

| 9 | 0 | 1 | 0 | 0 | 0 | 9 | 0 | 6 | 4 | 2 | 0 | 9 | 0 | 2 | 5 | 0 | 0 | 9 | 0 | 1 | 2 | 5 | 0 |

sign | integer | float

sign value : 5-9 = +ve , 0-4 = -ve
the decimal point position is decided
the encoding process, here it lies after
the first digit

$T_1 = +0.1000$ $T_2 = +0.6420$ $T_3 = +0.2500$ $T_4 = 0.1250$

Fig. (3-8) Chromosome construction Example

This method yields a faster encode/decode rate since the genes given are in real numbers.

3.4.2 Fitness function

The fitness function is the heart of an evolutionary optimization application. It is responsible of determining which solutions -within a population- are better for solving a particular problem. So the proper selection and development of the fitness function is essential for successful convergence in search results. In chapter 2, it was explained in details how to develop a state-space model for the SMIB power system used in this study. The developed fitness function is based on this model.

3.4.3 Fitness function evaluation

Basically, the fitness function evaluates how fit each chromosome is, and assigns a fitness value with the corresponding individual. In the present problem, the inputs are (T_1 to T_4) related to the PSS parameters set, and the output is chosen to be based on an error index reference to implement the modeling software into a fitness function. An m-file Matlab program (given in appendix [B]) is written to perform the state-space modeling of the system, performance index calculating and fitness function evaluation. This is shown in Fig. (3-9).

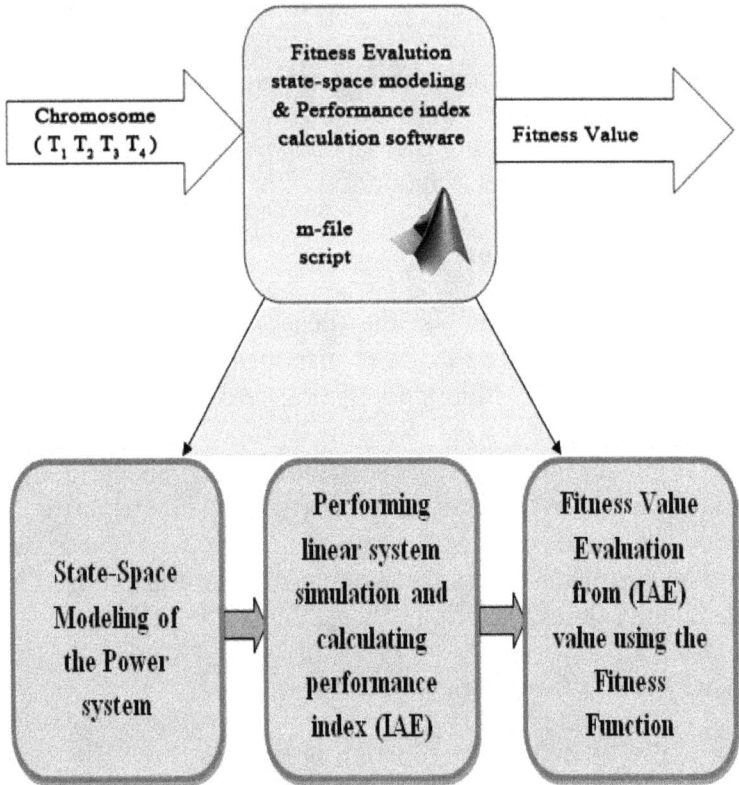

Fig. (3-9) Fitness function evaluation

Evaluation of the fitness value incorporates picking a chromosome from the current population and decoding this chromosome into a pheno-type mode that enables it to be used in the modeling software to perform time domain response. The fitness function has a multiobjective goal: First one is based on performing closed-loop analysis of the system step response to calculate a performance index **J** which given in the following equ. (3.4)

$$J = \frac{1}{N} \sum_{n=1}^{n=N} |\Delta \omega_r(n)| \qquad \qquad \dots\dots (3.4)$$

in order to minimize **J** or maximize $\frac{1}{J}$.

The second objective considered is a damping optimization rule for the damping factor **σ** and damping ratio **ζ**. Here, the eigen-value analysis is performed to find the lightly damped and undamped modes of oscillation. These modes will be shifted to the left of a wedge-shaped area whose boundaries are (σ_0, ζ_0) and these boundaries are not kept constant during the search process, as most previous studies did, but rather being optimized within the GA process. Fig. (3-10) shows a diagram explaining the mentioned boundaries. To assure the shifting of the locations of all lightly damped and undamped modes into the area defined by (σ_0, ζ_0), an instant rejection feature is embedded within the fitness evaluation section which assigns a low fitness value to these "bad individuals" to minimize the chance of reproduction in the next generation. This prepares for the selection step to be performed.

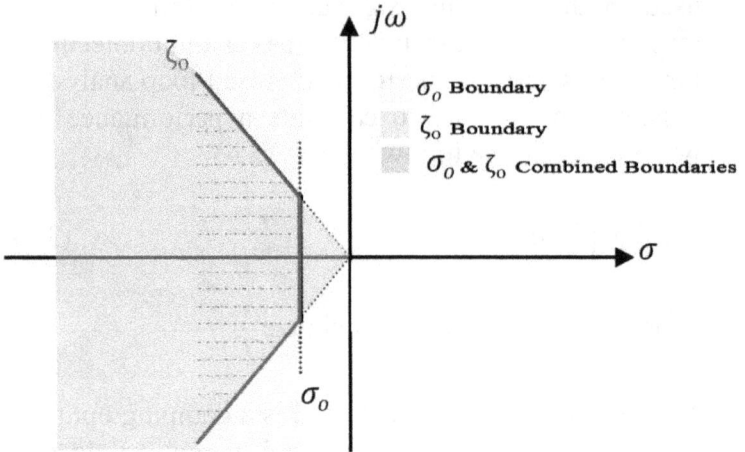

Fig. (3-10) Eigen-values boundary objective

3.4.4 Selection

The selection of individuals plays an important role in genetic algorithm. In this study, the Roulette-wheel selection method has been employed. Also, the Elitism feature is also utilized within the selection section. This assured passing the genetic information of the best individual to the next generation without being modified by other genetic operators. This Roulette-wheel technique gave the individual with a higher fitness value a larger portion in the collective probability values and therefore, has a higher chance of being copied into the mating pool while the individual with a smaller fitness will have a smaller probability of being copied into the mating pool [29].

3.4.5 Reproduction phase

Reproduction phase includes the following two operators:

1. **Crossover:** Two parents' chromosomes are chosen to be modified through this operator. The single-point crossover method was used in this study.

The modification to this method is made by taking two parents to produces a single child each time. Crossover doesn't occur with all individuals from the mating pool, a certain probability is imposed to the mating process. It is defined by crossover probability (P_c). So based on this probability, some parents are copied directly into the next population. While others may end up mating to reproduce new offspring [31]. Fig. (3-11) explain the single point crossover method used.

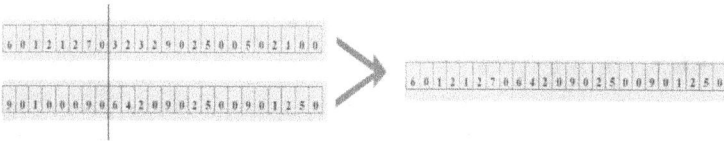

Fig. (3-11) Single-point crossover with single offspring

b. **Mutation:** According to the given mutation probability (P_m), some of the new offspring will go through a mutation. Here a random gene position is picked from the string and its value changes to another random generated value within the base 10 range (0-9). It is necessary to mention that the mutation probability is very low in comparison with the crossover probability. Fig. (3-12) explains the mutation method utilized in the study.

Fig. (3-12) Random point mutation

In both crossover and mutation operations, if the randomly chosen individual is the best fitness individual (Elite), it will be copied to the next population without any modification in order to maintain its genes quality and preserve information.

These two operators will define the next generation, and the operation continues with fitness evaluation and selection, etc. until a stopping criterion is reached.

3.4.6 Stopping criteria

This is defined in the GA by two goals whichever comes first:

1. The search process reaches the point where no noticeable improvement in the fitness value is encountered for a known number of generations, or
2. The generation counter reaches a defined maximum number, and here the last solution reached is taken and is considered the best possible optimization outcome.

3.4.7 The damping optimization

In the fitness evaluation section, it was mentioned that the values of damping factor and damping ratio (σ_o, ζ_o) are not kept fixed during the search, as all the previous works did [9], [15], [22], but rather optimized so as to have an optimum amount of damping based on system model and parameters. The process of damping optimization starts with initial values of damping factor and damping ratio (σ_{oi}, ζ_{oi}). These initial values depend on the eigen value analysis for the open-loop system. It is taken for the least damped pair of complex poles and the optimization begins the search for best values of (σ_o, ζ_o) with increments of ($\Delta\sigma$) and ($\Delta\zeta$). At each set of (σ_o, ζ_o), the GA process is run and this is repeated until the optimum values of (σ_o, ζ_o) found. The real strength of this approach is to have the damping properties set at optimum level or getting the minimum performance index **J**. It is based on the modeling parameters and the condition of the system model. This operation is continued to the point where the fitness is actually decreasing after two repeated tryouts. The area of search is defined by the initial set (σ_{oi}, ζ_{oi}) shown in Fig. (3-13).

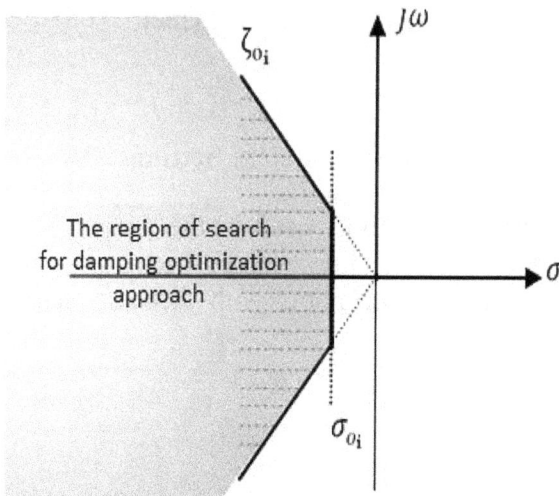

Fig. (3-13) Search space for (σ_o , ζ_o)

3.5 Search optimization flowchart

The whole search and optimization process is summarized using the flowchart of Fig. (3-14) which shows the main sections and steps used in this study and explains the mechanism of the search process and the basic cycle for the genetic algorithm and damping optimization method. The flowchart defines the main components of the search algorithm and it is totally implemented using Matlab environment and based on m-file programming, the written software is given in appendix (-B-). The flow chart shows the main steps in which the genetic algorithm implementation software executes, these steps are initializing and defining the GA , damping properties and SMIB data, performing the closed loop response to obtain the fitness value for each individual in the current generation, check the stopping criteria and optimal damping, if yes then the search has reached the desired solution and if not then the genetic operators are applied to produce the next generation and the search proceed in the same manner until a stopping criteria is met.

3.6 Genetic Algorithm Parameters Settings:

The GA parameters settings used in the present work are given in table (3-1), they have been chosen after performing several experiments with different settings of GA parameters.

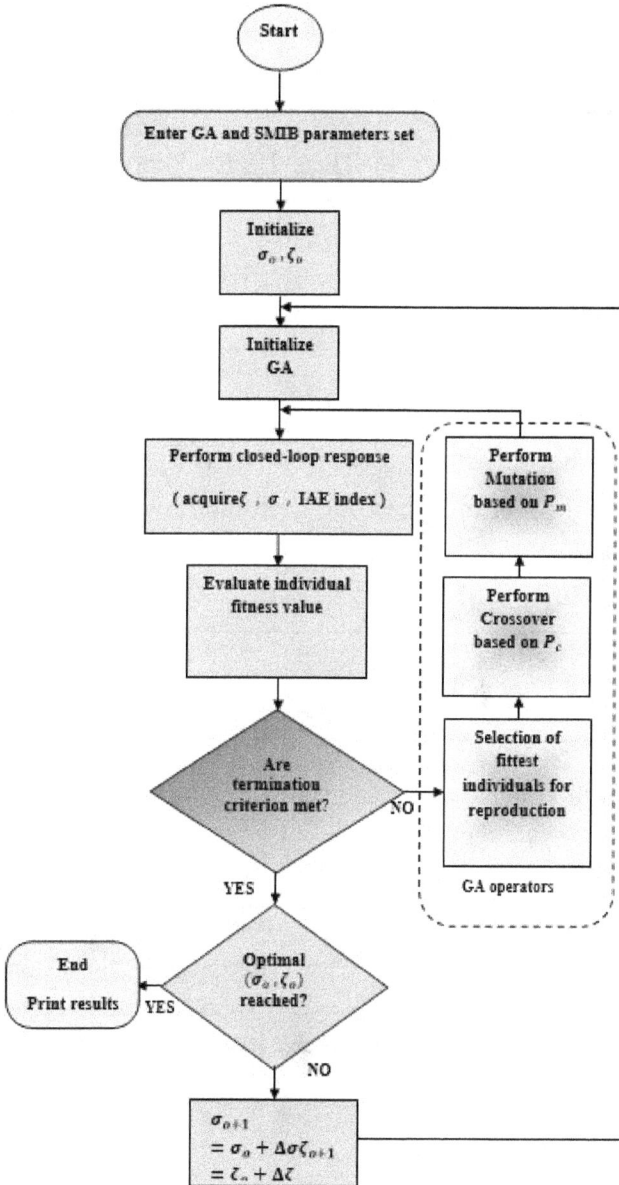

Fig. (3-14) Main GA process flowchart

Table (3-1) GA parameters setting

Encoding method	Base-10, Real coding with 24bit –per-string
Population size	75 chromosomes
Fitness goal	Maximization of fitness function
Selection method	Roulette-wheel
Elitism	ON
Crossover type & rate	Single-point crossover, ($P_c = 0.95$)
Mutation type &rate	Random gene mutation, ($P_m = 0.01$)
Stopping criteria	1. Maximum generations limit , or 2. No change in best fitness over (delta) generations
Maximum generation count	1000

Number of Gen. to be counted (delta)	80
Smallest improvement in best fitness (Epsilon)	0.001

3.7 Simulation Test Results

Detailed results for different system configurations is presented here, along with linear simulation results using the SMIB model subjected to different disturbances.

Table (3-2) provides details on the PSS setting and damping optimization values (σ_{op}, ζ_{op}) along with the best fitness value reached and PSS transfer function phase value ,given that the values of PSS gain block and washout time constant are **9.5** and **10** sec. respectively.

The approach taken to design a GA- PSS is evaluated using the (IAE) performance index. The value of performance index calculated as:

Performance index $= IAE = \frac{1}{N}\sum_1^N |e_i|$ ......(3.5)

where N is the total number of samples

and $e_i = \Delta\omega_r$

IAE is the integral of absolute error, is acquired and the amount of phase which is produced from the addition of the PSS to the power system model is also calculated.

The performance index is calculated for both approaches using the same situation which is subjecting the system to a step input (defined by the input matrix in the state-space) and calculating the value of IAE for duration of 15 seconds.

Table (3-2) PSS setting and damping optimization results

Design Method	PSS settings					Fitness	Damping optimization values	
	T_1	T_2	T_3	T_4	PSS Phase value	Best Fitness Value	Optimized damping factor (σ_o)	Optimized damping ratio (ζ_o)
GA-PSS (with damping optimization)	0.7376	0.1009	0.1539	0.2300	4.0492°	25.5461	-1.5000	0.2500

3.8 Linearized System Responses

The process of modeling and linearizing the power system and deriving of block diagram for the power system and the PSS was explained in details in chapter 2. Also, the traditional PSS design method was given with linear simulations. The system ratings and parameters are given in Appendix [A]. The GA- PSS performance was evaluated using the same linear system employed in traditional PSS simulations and under the same operation conditions. Block diagram of the linearized test system which was given in Fig. (2-16).

Linear simulations were performed to find the response of the system and observe the effect of the genetic algorithm based PSS. Fig. (3-15) shows the speed deviation response of the system when subjected to a step change at nominal machine loading of 0.9 pu active power and 0.3 pu reactive power.

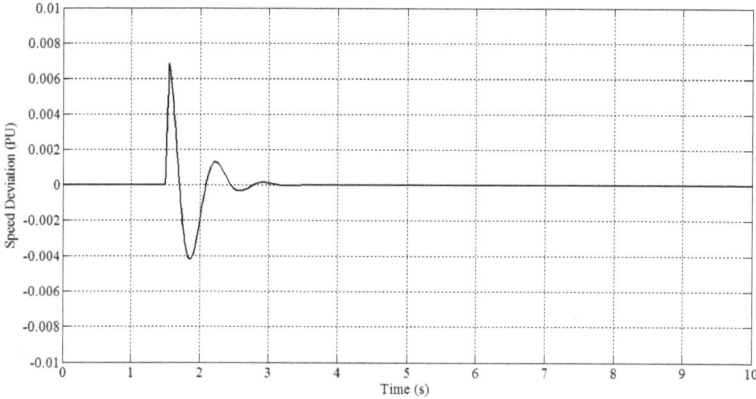

Fig. (3-15) Speed deviation response for system with GA- PSS

The step change disturbance was applied to the input at t = 1.5 s for a duration of 0.05 s. The response shows that the system remains stable after being subjected to the step change response disturbance at the mechanical input with better transient time characteristics (especially for the settling time) than the case of classic-PSS.

Also, rotor angle deviation response in Fig. (3-16) was taken from the linear simulation of the system at the same operating conditions.

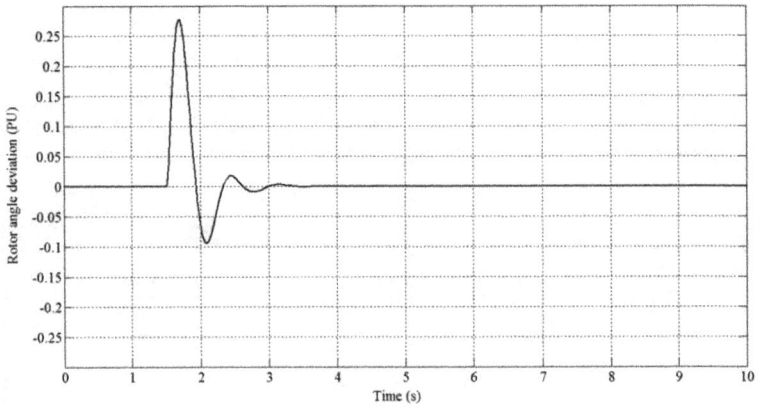

Fig. (3-16) Rotor angle deviation response for system with GA–PSS

CHAPTER FOUR

RESULTS, COMPARISON AND DISCUSSION

In this chapter, the linearized system given in Fig. (2-19) is used to develop a comparison of the different design approaches on the same system and under the same conditions. Also, a model of a single machine infinite bus system is developed in MATLAB®/ SimPowerSystems® environment. This simpower model is used to evaluate the performance of the proposed genetic algorithm based PSS (GA- PSS) along with the traditionally designed PSS (Classic- PSS).

Comparisons have been made between the results of simpower model and the linearized model. Different cases of loading conditions and disturbances are taken in order to clarify the behavior of the system with the PSS under various operating conditions.

4.1 Linearized SMIB System Model Simulation and Analysis

4.1.1 Eigen value results:

The linearized power system model with different configurations in terms of eigen-values and damping ratios is given in table (4-1). In order to develop a worthy comparison for different system configuration, the results were taken for the power system at the same operating conditions and ratings.

Table (4-1) SMIB system model eigen values and damping ratios for different configurations

System without AVR		System with AVR and without PSS		System with Classic PSS		System with GA-PSS		System with GA-PSS (with damp. Optimization)	
Eigen values	Damping ratio	Eigen values	Damping ratio	Eigen values	Damping ratio	Eigen values	Damping ratio	Eigen values	Damping ratio
$-0.1125 - j6.4041$	$+0.0176$	$+0.5061 - j7.2327$	-0.0698	-39.5075	$+1.0000$	-33.7792	$+1.0000$	-33.2911	$+1.0000$
$-0.1125 - j6.4041$	$+0.0176$	$+0.5061 - j7.2327$	-0.0698	$-19.5385 - j141.1803$	$+0.8090$	$-3.6289 - j7.9328$	$+0.4160$	-14.6865	$+1.0000$
-0.1875	$+1.0000$	-30.8755	$+1.0000$	$-19.5385 - j141.1803$	$+0.8090$	$-3.6289 - j7.9328$	$+0.4160$	$-3.6400 + j8.7795$	$+0.3270$
		-20.5400	$+1.0000$	$-1.0974 + j6.5462$	$+0.1650$	-6.7948	$+1.0000$	$-3.6400 - j8.7795$	$+0.3270$
				$-1.0974 - j6.5462$	$+0.1650$	-5.4184	$+1.0000$	$-3.6004 + j3.0429$	$+0.7640$
				-0.1605	$+1.0000$	-0.1605	$+1.0000$	$-3.6004 - j3.0429$	$+0.7640$
				-7.4544	$+1.0000$	-1.5414	$+1.0000$	-0.3035	$+1.0000$

In table (4-1), the given system configurations are:

1. **System without AVR**: This configuration is made by considering that the system is on manual excitation, so there is not any negative-phase is introduced as mentioned in section (2.2.2). Here only the effect of the first three states in the state-space model is apparent on the damping properties.
2. **System with AVR and without PSS**: In this configuration, an AVR (Modeled as a gain function with first order time delay derived from the voltage sensor) is added to the system. Therefore, the effect of adding negative phase value appears.
3. **System with Classic– PSS**: Here, the model is modified by the addition of a power system stabilizer which is designed using the traditional method of minimum phase design approach. It is shown that the order of the system is raised by three states due to the PSS model function.
4. **System with GA– PSS**: in this step, a PSS was also added to the system but it is designed using genetic algorithm approach instead of the classic method.
5. **System with GA– PSS (with damp. Optimization)**: In this step, the method of optimizing the damping properties (damping factor and ratio) is incorporated along with the genetic algorithm. This design takes into account the position of the boundary for the placement of the closed- loop eigen values in order to ensure the optimum value of damping.

Eigen value analysis:
(1) In the first case, where the system is running on manual excitation control (no AVR) and system model is of 3rd order. It is observed that the system is stable at these

conditions with a pair of complex conjugates and a single real eigenvalues lying at the left hand side of the s-complex plane.

(2) In the second case in which AVR unit was added to the system, now the system is of 4th order. The addition of the AVR caused the complex conjugate pair of eigen values to shift their position toward the right section of the s-plane, passing the stability boundary (jw-axis) leading the system into instability. This is also apparent from the negative value of the damping ratio related to this pair.

(3) In the third case, a PSS is added to the power system designed using the classic method of finding the minimum phase, system becomes of 7th order due to the addition of the PSS which raises the order by three states. Two pairs of complex conjugate and three real eigen values, all lying in the left hand side of s-plane appeared in the analysis. The positions of the eigen values along with the damping ratios suggest that the system has returned back to stable region.

(4) The fourth case shows the eigen value analysis for the power system that is equipped with a genetic algorithm based PSS. It was detected that one of the complex conjugate pairs was diminished into two real eigen values. The remaining pair stayed within the stable region of the s-plane. However, the boundary that keeps this pair within is not optimally decided as to yield an optimum damping properties (which is to optimally decide the certain values of damping ratio and damping factor).

(5) In the final case, An GA based PSS with damping optimization is applied to the power system. Here, two pairs of complex conjugates with three real eigen values appear in the analysis. In this design approach, it was found that the optimal boundary for the locations of the complex pairs, and thus the damping properties, is defined by $\sigma_o = -0.25$ and $\zeta_o = 0.25$ and this is confirmed by the eigen value analysis as it shows that the two pairs are kept within the area defined by the optimal values (σ_o and ζ_o).

4.1.2 Linearized system simulation responses:
In this section, linear simulations for the SMIB model with different configurations were performed using MATLAB®/Simulink® environment. The purpose of these simulations is to show the effect of different system components on the stability and to further evaluate the effectiveness of the proposed design in stabilizing the system. Table (4-2) shows the three operating conditions at which the simulations are carried out. The system is subjected to a full step change in loading for a period of (0.05) sec. and results are taken for these simulations and they are in terms of speed deviation ($\Delta\omega$) and rotor angle deviation ($\Delta\delta$) responses [42].

Table (4-2) Loading conditions during linear simulations

loading	P (PU)	Q (PU)	PF
Nominal	0.9	0.3	0.95
Heavy	1.0	0.75	0.8
Light	0.5	0.375	0.8

Linear simulations were performed on the system with different configurations while taking different operating conditions. These conditions are explained in details in table (4-2) which are nominal, light and heavy loading. Fig. (4-1) shows the simulation results for speed deviation response of four different system configurations (without AVR, without PSS, Classic- PSS design and GA-PSS design) taken at nominal loading.

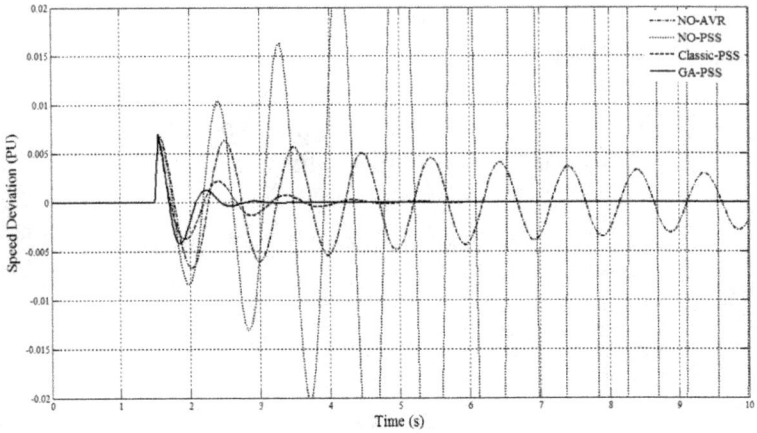

Fig. (4-1) Speed deviation response for four system configurations taken at nominal loading

Rotor angle deviation response of the system was also taken for the same case above. This response is shown in Fig. (4-2) and it is taken also at nominal system loading.

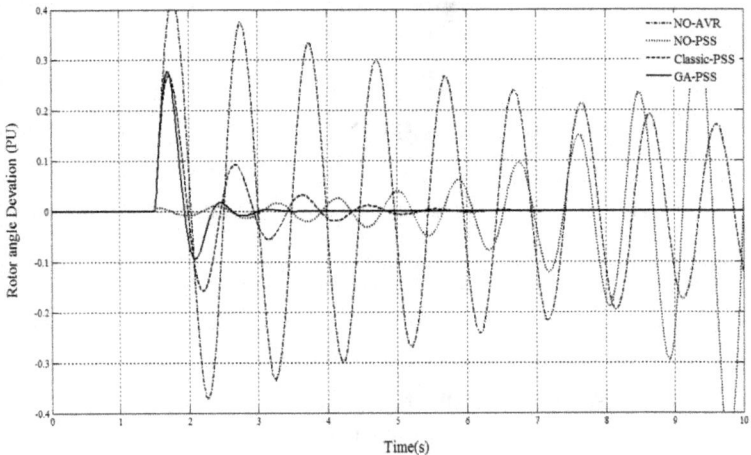

Fig. (4-2) Rotor angle deviation response for four system configuration taken at nominal loading

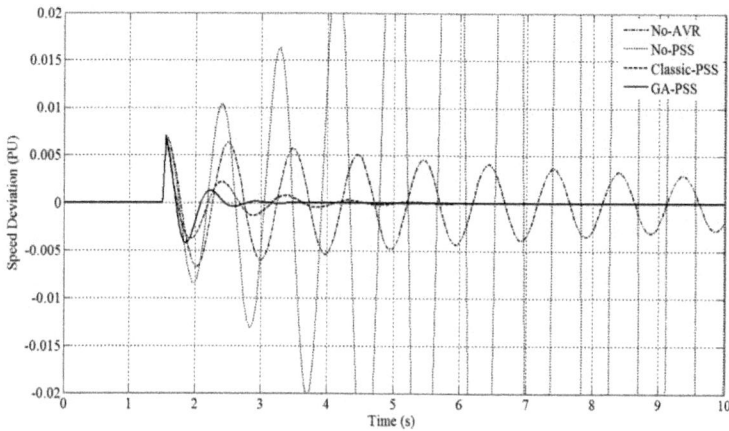

Fig. (4-3) Speed deviation response at heavy loading

Fig. (4-3) shows the speed deviation response for heavy loading with the same disturbance as in nominal loading condition. Rotor angle deviation response at heavy loading is shown in fig. (4-4) taken for four different system configurations.

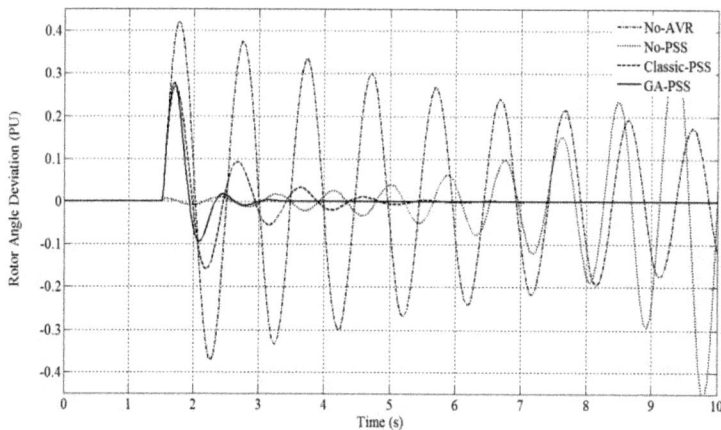

Fig. (4-4) Rotor angle deviation response at heavy loading

Fig. (4-5) shows the speed deviation response taken at light loading with the same applied external disturbance as in previous cases.

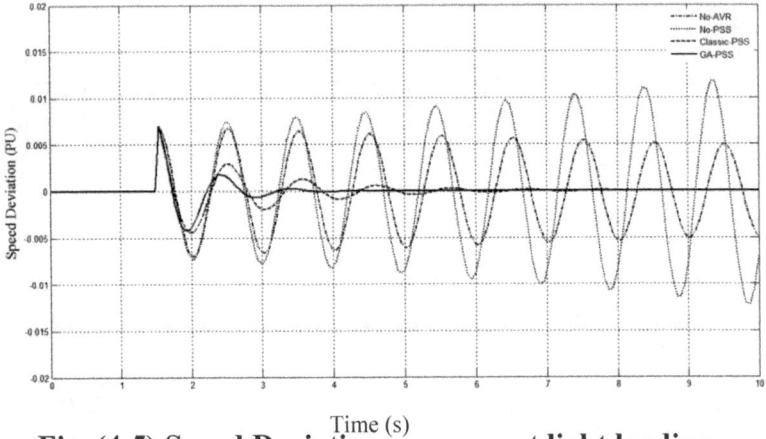

Fig. (4-5) Speed Deviation response at light loading

Rotor angle deviation response for the same light loading case with the same applied external disturbance is shown in Fig. (4-6).

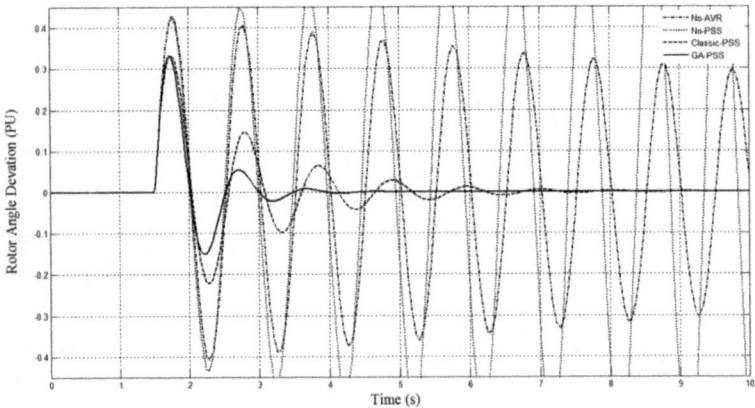

Fig. (4-6) Rotor angle deviation response at light damping

4.2 SMIB Model in SimPowerSystems® Environment

To evaluate the performance of the proposed genetic algorithm-based PSS design together with the traditional method of design which was anticipated in [1] and to make a valid assessment of both designs in stabilizing the power system at different operating conditions and loadings, a Simulink / PowerSystemsBlockset model was constructed. This Simulink model was built with data given in app. [A]which have been taken from [1] in order to make a suitable comparison between the two design methods as this model is the nearest representation to a practical power system. The power system model built up here consists of a single generating unit connected to a transmission line via step up transformer and to an infinite bus. The synchronous generator has a control block which contains all control devices (automatic voltage regulator, governor and power system stabilizer). Fig. (4-7) shows the SMIB power system model considered. The model is based on the MATLAB® Simulink®/SimPowerSystems® software which is a modern design and analysis tool that allowing the designer to rapidly build and simulate complex power system models and replicate their response behaviors allowing the designer to build the power system model in a convenient way with the completed circuit topology. The analysis of the model shows the interaction with other mechanical, thermal and control restraints. This is possible due to the simulation interface with the wide-ranging Simulink modeling library.

The main parts of the power system considered in this model are:

a. **Synchronous generator**: The generating unit employed here is a 3-phase salient pole synchronous machine modeled in the dq- rotor reference frame in per unit system. It is represented by a sixth-order state space model and it takes into account the dynamics of the stator, field and damper windings

This machine is equipped with a control and instrumentation block which contains all the necessary controlling devices such as (excitation system, governor and power system stabilizer blocks) along with different measurements busses and ports.

b. **Step-up transformer**: This is a 3-phase transformer (two windings, using 3 single phase transformers) which raises the generator voltage level to the grid high tension level.

c. **Transmission line circuit**: This part is an OHTL circuit in a PI section configuration which implements a 3-phase transmission line with parameters defined in the PI section.

d. **Voltage source**: This block is a 3-phase balanced voltage source with internal R-L impedance which is equivalent to an infinite bus.

Other parts and blocks in the power system model considered are used in setting the initial and reference quantities for different parts of the circuit, initializing the power system to a steady-state point, taking various measurements and displays for different parameters applying a mixture of disturbances to the system and monitoring the state of the system and the status of synchronism. Both types of power system stabilizers (Classic- PSS and GA- PSS) are implemented in this power system.

Fig. (4-7) SMIB power system model in Simulink®/SimPowerSystem blockset

The model in Fig. (4-7) was initialized to the point when the system reached steady state condition. Then tests were executed in order to assess the performance and validity of different design approaches (traditional and genetic based ones).

4.2.1 Traditional compensator.

In this section, the performance of the classic PSS is investigated by running the system and subjecting it to external disturbances. The results were taken from both models. The linearized block diagram model shown in Fig. (2-19) and the simpower model shown in Fig. (4-7). In both models, the loading conditions and the external disturbance were the same. These loading conditions are given in table (4-3) and the disturbance is a step change for a period of 0.05s.at the system mechanical input.

Table (4-3) Loading condition for simulation tests

	Active power - Pe– (per unit)	Reactive power - Qe- (per unit)
Heavy loading	1.0	0.75
Medium loading	0.7	0.25
Light loading	0.5	0.12

The first case is shown in Fig.(4-8) which is a speed deviation response for the linearized model without PSS at nominal loading with subjected to a step change for a 0.05sec.at t= 51.5 sec.

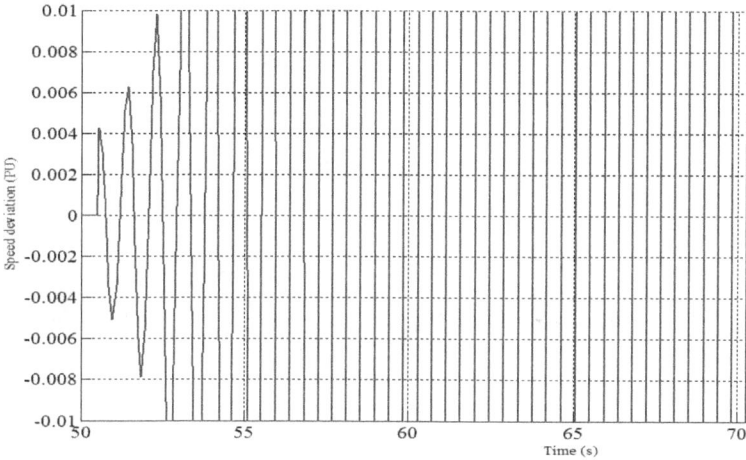

Fig. (4-8) linearized model speed deviation response of the system without PSS

Without the application of a PSS, it is obvious from Fig. (4-8) that the system becomes unstable at nominal loading condition when subjected to disturbances. Now taking the traditional design results and applying them to both models, we obtain rotor speed deviations response at different loading conditions as in table (4-3). Fig. (4-9) shows the response of the linear model at heavy loading with classic PSS. Fig. (4-10) shows the speed deviation response of the simpower model at heavy loading with classic PSS.

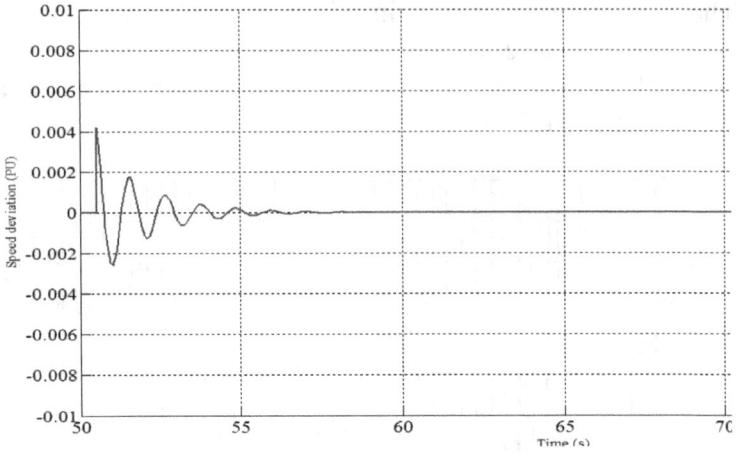

Fig. (4-9) System speed deviation response for linearized model at heavy loading with Classic- PSS

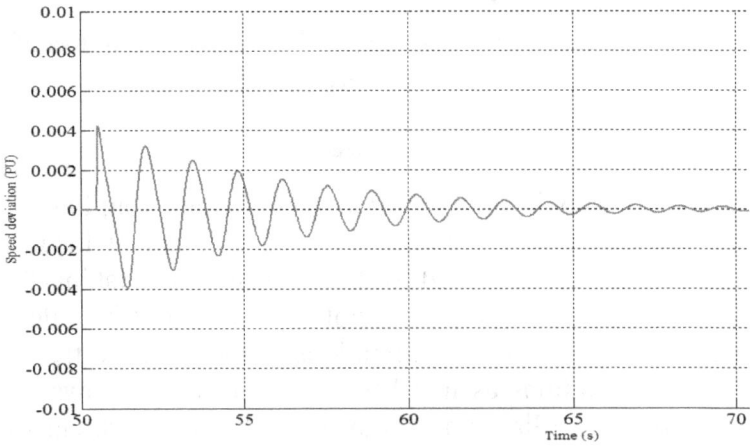

Fig. (4-10) System speed deviation response for simpower model at heavy loading with Classic- PSS

At light loading conditions, the linearized model was subjected to the external disturbance and speed deviation response, shown in Fig. (4-11).

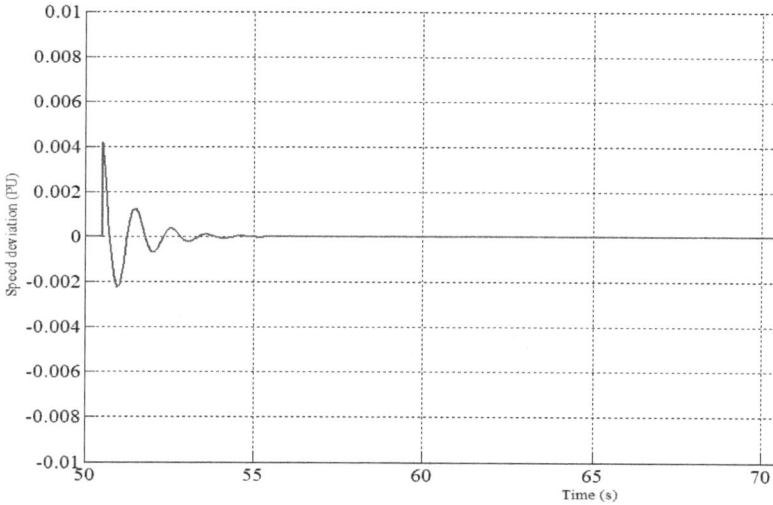

Fig. (4-11) System speed deviation response for linearized model at light loading with Classic- PSS

Fig. (4-12) shows the speed deviation response for the simpower model at light loading and same external disturbance.

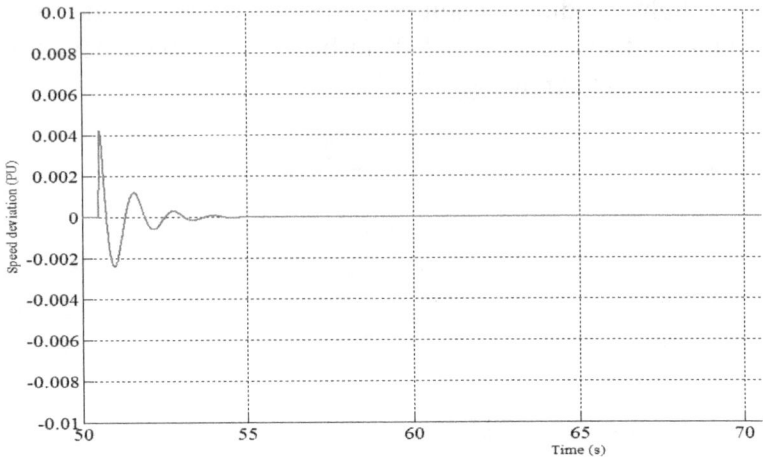

Fig. (4-12) System speed deviation response for simpower model at light loading with Classic- PSS

It could be seen at heavy loading that both models are not the same but they have close results, while at light loading the results are identical in terms of maximum deviation and settling time.(ts =**4.5s**. at 2% criterion and max.d.=**0.004**).

4.2.2 Genetic-algorithm based compensator.

The GA-PSS performance was evaluated at the loading conditions given in table (4-3) by applying the same external disturbance taken in (4.2.1) and monitoring the rotor speed deviation response afterwards. First the linearized model with GA-PSS was tested at heavy loading condition and the speed deviation response is given in Fig. (4-13).

Fig. (4-13) System speed deviation response at heavy loading for linearized model with GA- PSS

The simpower model with GA-PSS was also tested at heavy loading with the same applied disturbance and the speed deviation response is given in Fig. (4-14)

Fig. (4-14) System speed deviation response at heavy loading for simpower model with GA- PSS

Now for light loading, the linearized model was tested and speed deviation response is given in Fig. (4-15)

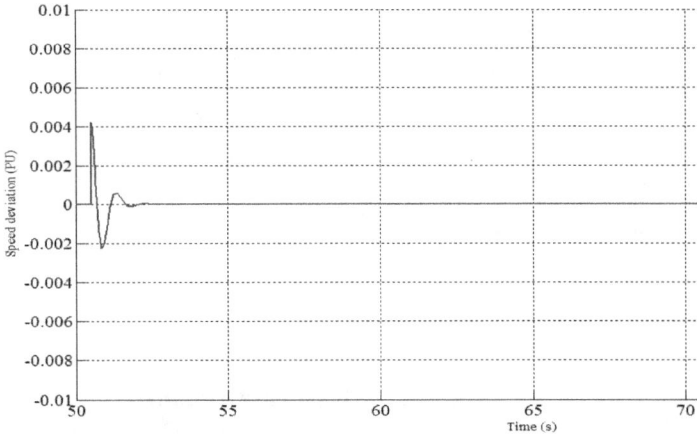

Fig. (4-15) System speed deviation response at light loading for linearized model with GA- PSS

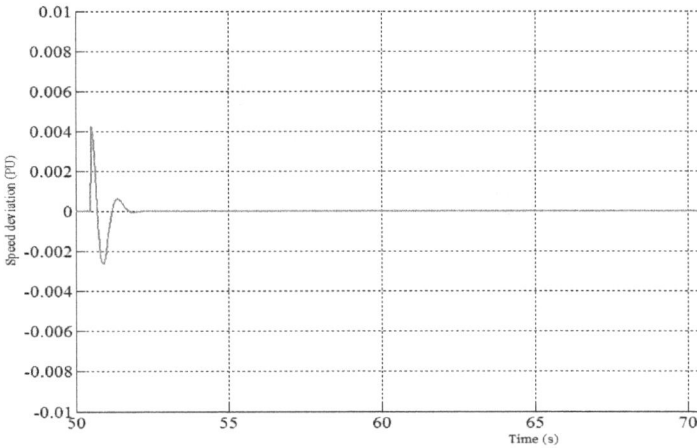

Fig. (4-16) System speed deviation response at light loading for simpower model with GA- PSS

For the simpower model at light loading, the speed deviation response is given in Fig. (4-16)

Comparing both models at heavy and light loading, it is evident that the results are very close in maximum deviation and settling time. (At light loading, ts = **1.3**s. at 2% criterion and max.d. = **0.004**).

To have a clearer perspective on these cases, the speed deviation of both designs at each loading conditions was plotted on the same graph for both models. Fig. (4-17) shows the speed deviation response of the linear model at light loading for both design approaches and Fig. (4-18) shows the system speed deviation response for linearized model at heavy loading.

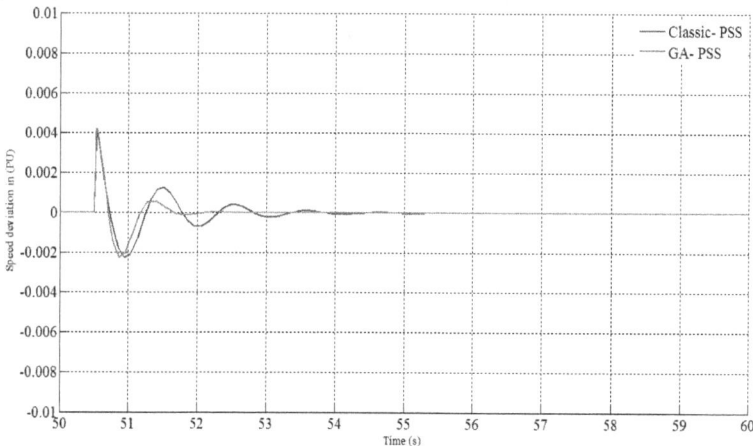

Fig. (4-17) System speed deviation response for linearized model at light loading with Classic- PSS & GA- PSS

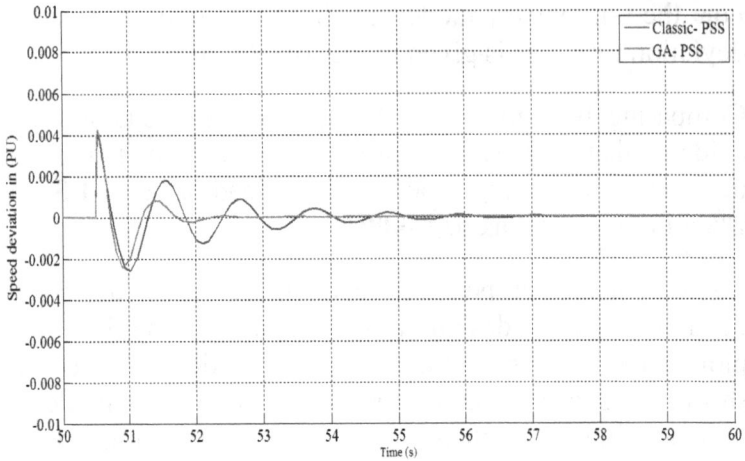

Fig. (4-18) System speed deviation response at heavy loading for linearized model with Classic- PSS & GA- PSS

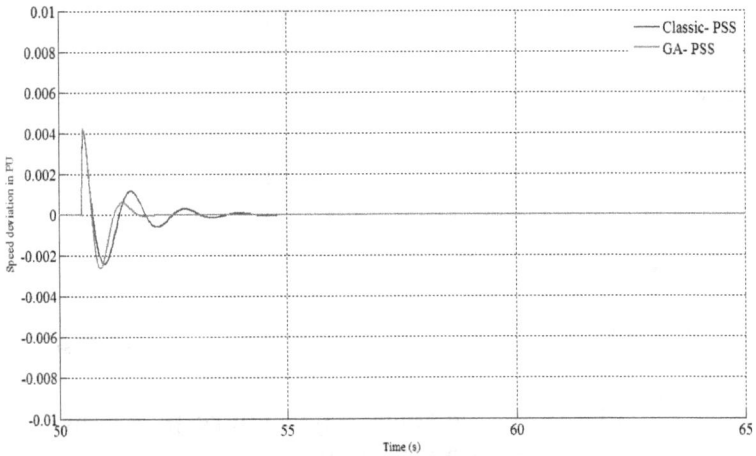

Fig. (4-19) System speed deviation response for simpower model at light loading with Classic- PSS & GA- PSS

For the linear model at heavy loading, ts=**4.2**s for Classic-PSS while for GA-PSS, ts=**1.3** s.)For simpower model, the speed deviation response at light loading is given in Fig. (4.-19). While for heavy loading, the speed deviation response is given in Fig. (4-20).

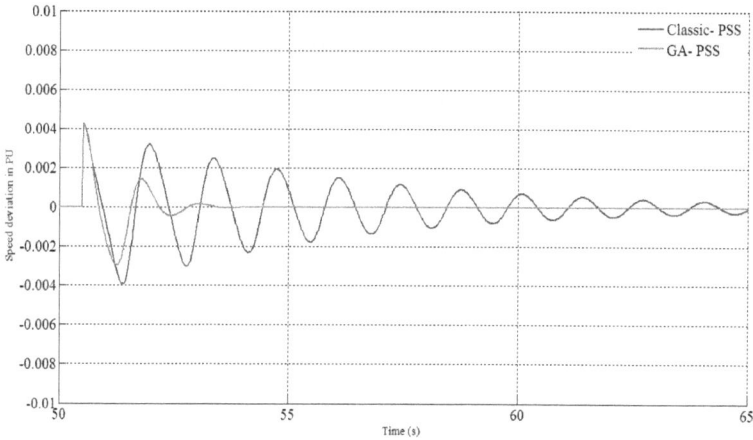

Fig. (4-20) System speed deviation response for simpower model at heavy loading with Classic- PSS & GA- PSS

These results for simpower model show the effectiveness of the GA-PSS over the Classic-PSS. At heavy loading, the settling time for the system with Classic-PSS was **10.5** s while for GA-PSS it was **2.5** s.

4.3 Simpower Model Simulation at Different Loadings

To have a better understanding on the behavior of the power system for different operation conditions, the simpower model was reconfigured in order to be subjected to different disturbances. The rotor speed ω_r, the active power Pe and the reactive power Qe, all in per unit, were monitored during the application of different disturbances and their responses are given in the following graphs. The first case was for the machine, supplying active power **Pe= 0.8** per unit at reactive power **Qe=0.32**per unit. A sudden raise of 20% in power demand was applied to the machine at t=45 s. The speed response is given in Fig. (4-21).

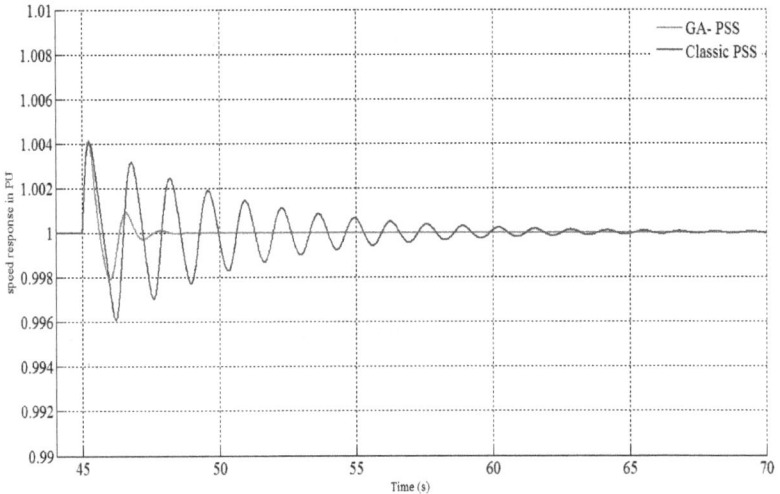

Fig. (4-21) System speed response with Classic- PSS & GA- PSS for ΔPe=+20% disturbance at Pe=0.8 per unit

and the active power Pe response is given fig.(4-22).

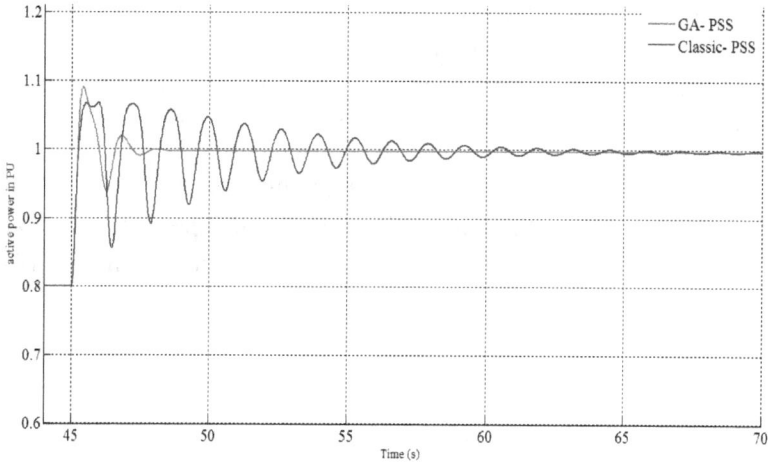

Fig. (4-22) System active power response with Classic-PSS & GA- PSS for ΔPe=+20% disturbance at Pe=0.8 per unit

The reactive power response is given in Fig. (4-23) below.

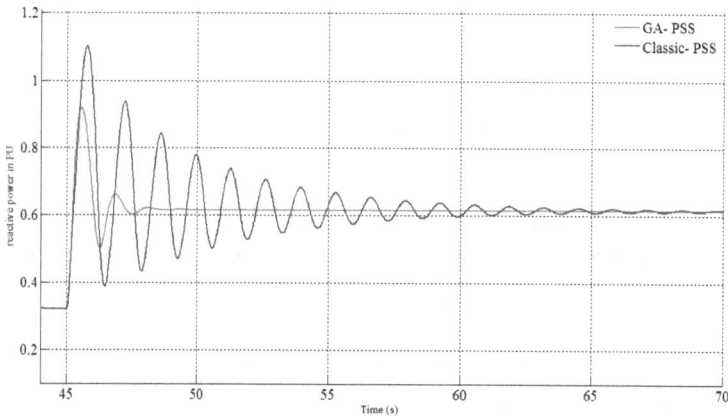

Fig. (4-23) System reactive power response with Classic-PSS & GA- PSS for ΔPe=+20% disturbance at Pe=0.8 per unit

These figures demonstrate the role of the proposed GA-PSS in stabilizing the machine active and reactive power in addition to the rotor speed during sudden changes in load demand up to the full machine loading capability.

The second case is applying a change of ΔPe= +20% at t=45s.thenΔPe=-40% at t=80s.when**Pe=0.8** per unit and **Qe** was **0.32** per unit. The speed response is shown in Fig. (4-24).

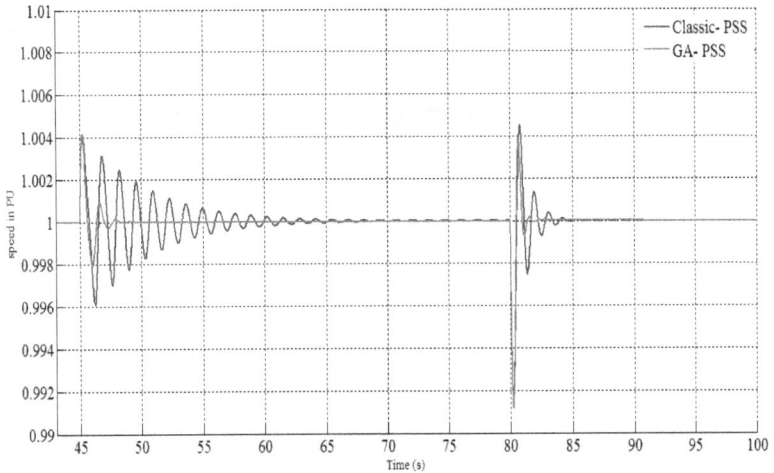

Fig. (4-24) System speed response with Classic- PSS & GA- PSS for ΔPe = +20% at t=45s. and -40% at t=80s. at Pe=0.8 per unit

The active power response at these operating conditions is shown in Fig. (4-25).

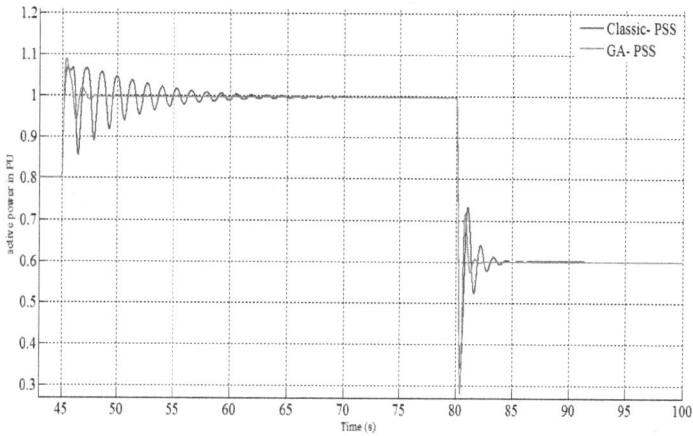

Fig. (4-25) System active power response with Classic-PSS & GA- PSS for ΔPe = +20% at t=45s. and -40% at t=80s. at Pe=0.8 per unit

Also the reactive power response is shown in Fig. (4-26) below

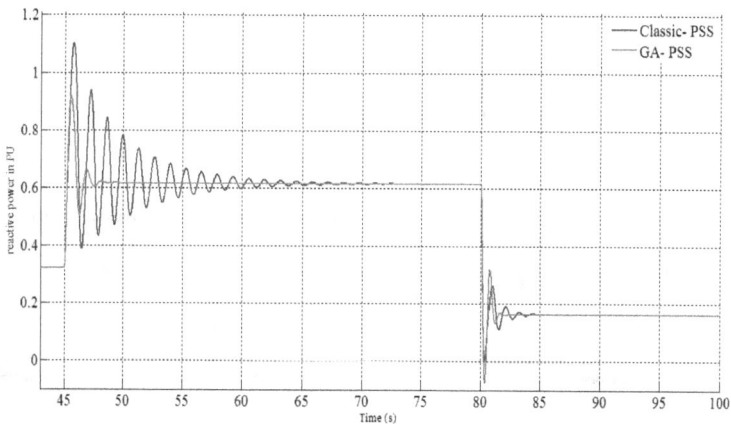

Fig. (4-26)System reactive power response with Classic-PSS & GA- PSS for ΔPe = +20% at t=45s. and -40% at t=80s. at Pe=0.8 per unit

It is clear from figures (4-24) to (4-26) that the proposed design is better in providing supplementary damping torque to the machine at wide range of loading and operating conditions.

The efficiency of the design is evident from the active and reactive power responses. Table (4-4) gives detail on the responses of the active and reactive power for $\Delta P_e = 20\%$ at loading of $P_e = 0.8$ per unit.

Table (4-4) Settling time for active power at $\Delta P_e = 20\%$

	Classic- PSS	GA- PSS
Settling time (sec.) for Active Power	9.5	1.8
Settling time (sec.) for P_e	20.5	3.3
Peak overshoot (%) for Q_e	83	51

Taking the third case of disturbance, the machine active power and thus loading was kept fixed, but the field voltage was subjected to a slight change at steady-state operation.

This gives an insight to the system's behavior when sudden changes in field voltage happens, also in helps finding the maximum change in these parameters that the system could handle without going out of synchronism .

1. When the machine was supplying active power **Pe = 0.8**per unit and reactive power **Qe = 0.32**per unit and the field voltage was changed by -5% at t=45 s. the results were as follows: Fig. (4-27), Fig. (4-28) and Fig. (4-29) show the speed, active power and reactive power responses of this case, respectively.

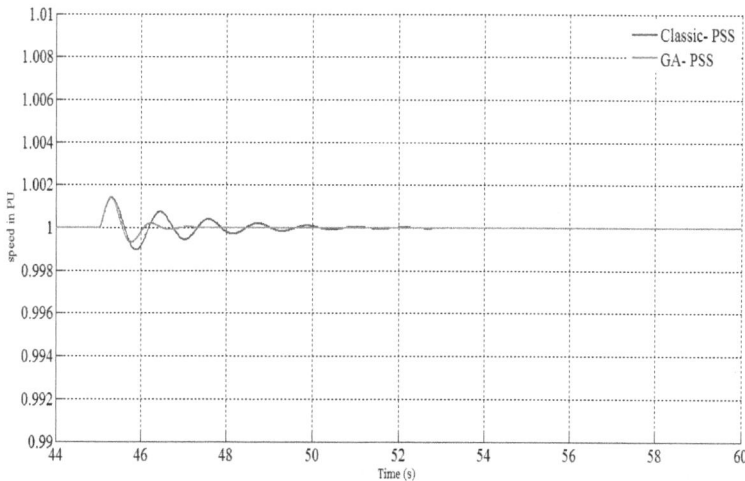

Fig. (4-27) System speed response with Classic- PSS & GA- PSS for ΔVf = -5% at t=45s at Pe = 0.8 per unit

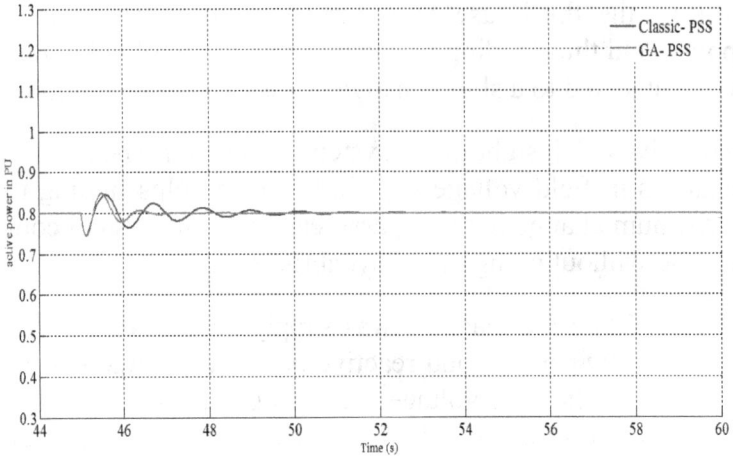

Fig. (4-28) System active power response with Classic-PSS & GA- PSS for ΔVf = -5% at t=45s at Pe = 0.8 per unit

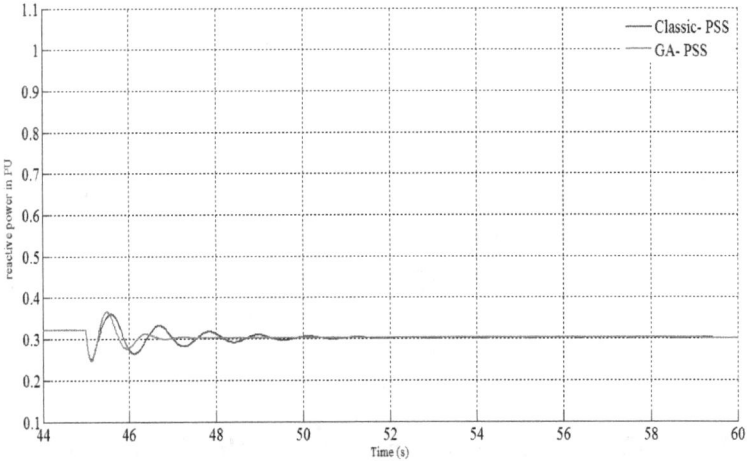

Fig. (4.29) System reactive power response with Classic-PSS & GA- PSS for ΔVf = -5% at t=45s at Pe = 0.8 per unit

The system seems to be in stable zone when subjected to the mentioned disturbance. Both designs gave acceptable performances even though the GA- PSS has better damping characteristics compared to the Classic- PSS.

2. In this case, the machine was supplying 1.0 per unit active power and 0.61 per unit reactive power. Then a disturbance of $\Delta Vf = -5\%$ at t=45s was applied to the system. The response of both designs (Classic-PSS and GA- PSS) for speed is shown in Fig. (4-30), active power response is shown in Fig. (4-31) and reactive power response is shown in Fig. (4-32).

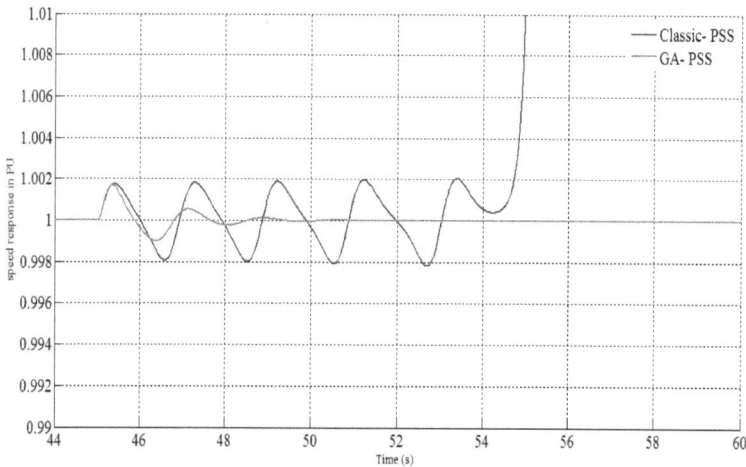

Fig. (4-30) System speed response with Classic- PSS & GA- PSS for $\Delta Vf = -5\%$ at t=45s at Pe = 1 per unit

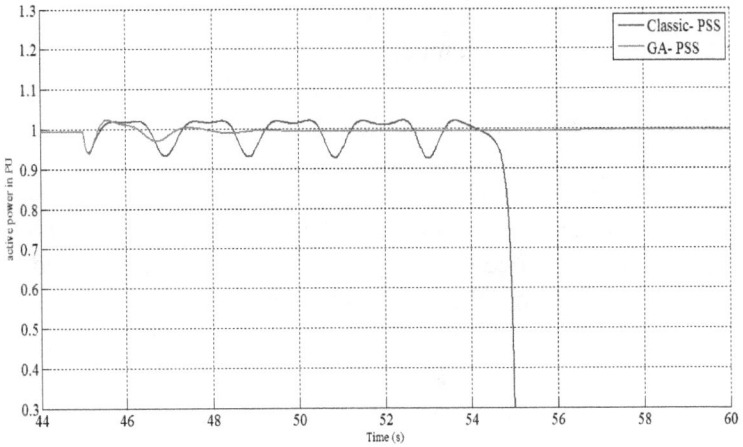

Fig. (4-31) System active power response with Classic-PSS & GA- PSS for ΔVf = -5% at t=45s at Pe = 1 per unit

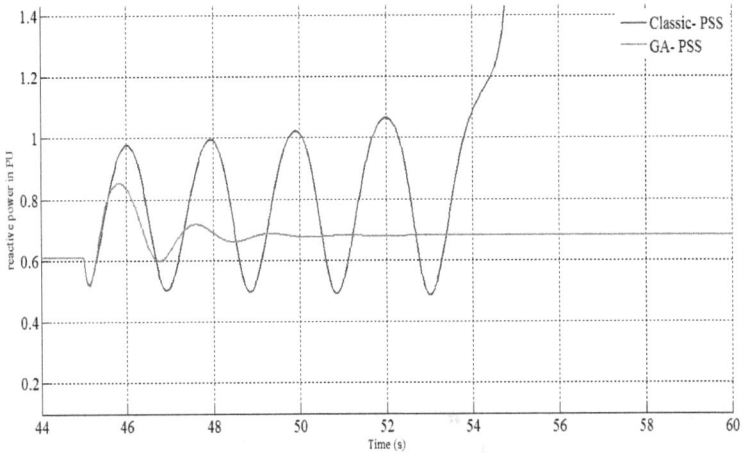

Fig. (4-32) System reactive power response with Classic-PSS & GA- PSS for ΔVf = -5% at t=45s at Pe = 1 per unit

It is evident that when the system was operating with Classic-PSS, it was not able to hold its synchronism with the applied disturbance in field voltage. This is clear in the speed, active power and reactive power responses. While with the proposed GA- PSS design, the controller was able to provide a sufficient damping component and restores the system to steady-state conditions.

These cases clarify the effectiveness of the proposed GA-PSS controller design in stabilizing the system over a relatively wide range of operating conditions compared to the traditionally designed controller.

To conclude the evaluation, it is better to give a parallel comparison for the results of both system test (linearized model and simpower model) at the same operating condition and disturbance.

The loading condition taken here is when the machine was supplying a steady-state load of active power **Pe = 0.9** per unit and reactive power **Qe = 0.44** per unit. This system was subjected to a step change for a period of t=0.05s. at its input.

Table (4-5) gives the results in terms of maximum deviation (max.d.) and settling time (ts) for speed deviation response from the linearized model and simpower model for open-loop, Classic- PSS and GA- PSS cases.

Table (4-5) Linear model speed deviation comparison

		Open-loop	Classic- PSS	GA- PSS
Maximum deviation	Linearized Model	-	0.0042	0.0040
	Simpower model	-	0.0042	0.0040
Settling time (sec.)	Linearized model	-	4.0	1.39
	Simpower model	-	8.2	2.1

The enhancement in settling time is evident from the table. In terms of settling time, the GA- PSS has reduced the settling time of the linear model from ts = **4.0** to ts = **1.39** seconds comparing to the Classic- PSS. In the simpower model, the improvement is further assured as the GA- PSS shortened the settling time from ts = **8.2** to ts = **2.1** seconds.

It should be mentioned that the difference in the results for linear model compared with the simpower model is due to the dissimilarity in the degree of details between the two models and the implementation of the various component of both models. Some of the differences between the two models are:

1. The linear SMIB model is linearized around a nominal operating point while the simpower model is non-linear (much like a practical system).
2. In the linear model, the mechanical input is considered constant without regulation where in the simpower model the input is regulated with a governor controller.
3. The linear model was developed considering that the machine is without damper windings, but the simpower model, the damper windings are available and taken into account.

Also, it was not possible to initialize and operate the power system in open-loop (without a PSS) configuration under these operating conditions.

On the other hand, the simpower model and linear model results are close to each other to a large degree. This leads us to say that the linear model is acceptable for analysis and design of power system and it may be a reliable model for such a purpose of study.

CHAPTER FIVE

CONCLUSION & FUTURE WORKS

5.1 Conclusion

The problem of power system stabilizer design has investigated in this book. A new approach aiming to solve the problem of power system stabilizer design has been introduced which is based on genetic algorithm with damping properties optimization method.

Small signal stability of the single machine infinite bus system has been investigated. The mathematical modeling and analysis of the system have been performed which resulted in forming a state-space linearized model based on solving system nonlinear equations around a certain operating point defined by active and reactive power values. The derivation of the state-space model made it possible to analyze the response of the open-loop system prior to adding the AVR unit and after the addition of the AVR unit to the system. This analysis demonstrated the negative effect of the AVR unit on system stability, given in Fig. (2-13) when the machine was operating on a stressed line and step disturbance was applied to the system input. A solution has been introduced in the form of power system stabilizer. The traditional design method of the PSS and system simulations after the addition of the PSS were carried out. The proposed genetic algorithm based design approach was applied to design the PSS. A novel feature was added to the search process such that it finds the optimum position of damping factor (σ) and damping ratio

(ζ) which in turn should ensure the optimum placement of the closed loop eigen values in addition to the PSS parameters optimization. Results of this method give in Figs. (3-16) to (3-17) were used in linear system simulations using the Simulink® environment to show system response during disturbances and system upsets.

Non-linear system simulations based on SimPowerSystems® environment (simpower model) for the given SMIB system were also adopted. A comparison of the results between linear model and simpower model for each design method (traditional and GA-based) was performed, given in Figs. (4-17) to (4-20). It has been shown that both model gave very close results in terms of system response due to subjected disturbances at heavy loading condition, with a slightly different response at heavy loading due to reasons given in chapter 4. In addition, a comparison of both design approaches for different operating conditions and disturbance using simpower model was carried out. The effectiveness of the GA-based design is clear in terms of settling time, peak overshoot and maximum deviation, as shown in tables (4-4) and (4-5). Also, the robustness of the proposed approach was evident when the machine was supplying a full load and the field voltage was lowered by 5%, shown in Fig. (4-30), the traditional design was not able to stabilize the machine after the applied disturbance, whereas the GA-based design was able to provide sufficient supplementary damping for the generator to hold its synchronism with the rest of the system.

The previous conclusion reveals the superiority of the proposed GA-based design over the traditional phase compensation based design in stabilizing the same tested

SMIB system in different loading conditions and applied disturbances.

5.2 Future Work

- Using evolutionary programming techniques to design the PSS like PSO, etc. to reduce the computation time and ensuring search convergence.
- Implementing the results practically.
- Using a DSP kit, FPGA Kit or other in realizing the PSS practically.
- Examining the GA based PSS design for multi-machine, multi-plant or multi-machine, multi PSS systems.

REFERENCES

[1]   P. Kundur, "Power system stability and control", McGraw-Hill, Inc. , 1994.

[2]   P. M. Anderson, A. A. Fouad, "Power system control and stability", 2nd edition, 2003.

[3]   F. P. DeMello, C. Concordia, "Concepts of synchronous machine stability as affected by excitation control", IEEE transactions on power apparatus and systems, vol. pas-88, no. 4, April 1969.

[4]   Adrian Andreoiu, " On power system stabilizers: genetic algorithm based tuning and economic worth as ancillary services ", PhD dissertation, Chalmers university of technology, Göteborg, Sweden, 2004.

[5]   Jan Machowski, Janusz W. Bialek, James R. Bumby, " Power system dynamics, stability and control ", 2nd edition, John Wiley & Sons, Ltd, 2008.

[6]   Graham Rogers, " Power system oscillations ", Kluwer academic publishers, 2000.

[7]   Saeid Kyanzadeh, Malihe M. Farsangi, Hossein Nezamabadi-pour, and Kwang Y. Lee, " Design of power system stabilizer using immune algorithm ", The 14th International conference on intelligent system applications to power systems, ISAP, Taiwan, November 2007.

[8]   Thomas Wiese, " Global optimization algorithms - theory and application ", Thomas Wiese electronic publications, Ānhuī, China, 2009.

[9]   Y. L. Abdel-Magid, M. A. Abido, S. Al-Baiyat and A. H. Mantawy, " Simultaneous stabilization of multimachine power systems via genetic algorithms ",

IEEE transactions on power systems, Vol.14, No.4, November 1999.

[10] E. V. Larsen, D. A. Swann, " Applying power system stabilizers Part I-III ", IEEE transactions on power apparatus and systems, Vol. PAS-100, No. 6 June 1981.

[11] Maslennikov, V. A., Ustinov, S. M., " Method and software for coordinated tuning of power system regulators ", IEEE transactions on power systems, Vol.12, Issue 4, pp. 1419-1424, November 1997.

[12] M. K. El-Sherbiny, G. El-Saady and E. A. Ibrahim, " Efficient incremental fuzzy logic for power system stabilization ", Machines and power systems, Vol. 25, pp. 429-441, 1997.

[13] M. M. Salem, A. M. Zaki, O. A. Mahgoub, E. Abu El-Zahab and O. P. Malik, " Experimental verification of a generating unit excitation neuro-controller ", IEEE power engineering society winter meeting Vol. 1, pp. 585-590, 2000.

[14] M. A. Abido, " Parameter optimization of multimachine power system stabilizers using genetic local search ", Electrical power and energy systems 23, pp. 785-794, 2001.

[15] Y. L. Abdel-Magid, M. A. Abido, " Optimal multiobjective design of robust power system stabilizers using genetic algorithms ", IEEE transactions on power systems, VOL. 18, NO. 3, August 2003.

[16] Naji A. Al-Musabi, Zakariya M. Al-Hamouz and Hussain N. Al- Duwaish, " Design of variable structure stabilizer for a nonlinear model of SMIB system : particle swarm approach ", WESAS transactions on

power systems, Issue 2, Vol. 1 pp. 311-316, February 2006.

[17]   G. Y. Yang, Y. Mishra, Z. Y. Dong and K. P. Wong, "Optimal power system stabilizer tuning in multi-machine system via an improved differential evolution ", Proceedings of the 17th world congress, the international federation of automatic control, Seoul, Korea, July 6-11, 2008.

[18]   Carlos A. C. Coello, Gary B. Lamont and David A. Van Veldhuisen, " evolutionary algorithms for solving multi-objective problems ", Springer science + business media, LLC, 2007.

[19]   Jayapal R., Dr. J. K. Mendiratta, "H∞ controller design for a SMIB based pss model 1.1 " Journal of theoretical   and   applied   information technology, Vol.11, No.1, January 2010

[20]   Eslami M., Shareef H. and Mohamed A., "Coordinated design of PSS and TCSC controller for power system stability improvement ", ," In Proceedings to the 2010 IPEC, Singapore, pp. 433-438, 2010.

[21]   R. Shivakumar, Dr. R. L. Lakshmipathi, " Implementation of an innovative bio inspired GA and PSO algorithm for controller design considering steam GT dynamics ", IJCSI International journal of computer science issues, Vol. 7, Issue 1, No. 3, January 2010.

[22]   Dr. Akram F. Bati, "Damping of power systems oscillations by using genetic algorithm-based optimal controller ", 1st International conference on energy, power and control (EPC-IQ), 2010.

[23]   O. A. Hussain, " Implementation of fuzzy logic using microcontroller for power system stabilizer "

M.Sc. thesis, college of engineering, university of Mosul, 2011.

[24] M. A. Abido, "Robust design of multimachine power system stabilizers using simulated annealing", IEEE transactions on energy conversion, Vol. 15, No. 3, September 2000.

[25] Mohamed Zellagui, "Robust power system stabilizer design using genetic local search technique for single machine connected to an infinite bus", International journal of signal system control and engineering application 1(3): 188-194, 2008.

[26] PG&E staff, "Power system stabilizer for generation entities ", PG&E interconnection handbooks, January 2010.

[27] Alex J. Champandard, " Genetic algorithms warehouse ", AI-depot website, available at : http://geneticalgorithms.ai-depot.com/, 2007.

[28] John H. Holland, " Adaptation in natural and artificial systems ", MIT press, 1992.

[29] Darrell Whitley, " A genetic algorithm tutorial ",computer science department, Colorado state university, 1993.

[30] John R. Koza, " Genetic programming: on the programming of computers by means of natural selection " 6th edition, MIT press, 1998.

[31] Tom V. Mathew, " genetic algorithm ", Indian institute of technology Bombay, Mumbai, India, 2005.

[32] Alden H. Wright, " Genetic algorithms for real parameter optimization ", Department of computer science, university of Montana, 1991.

[33] Randy L. Haupt ,Sue Ellen Haupt, " Practical genetic algorithms ", 2nd edition, John Wiley & Sons, Inc., 2004.

[34]   Kevin M. Passino, "Biomimicry for optimization, control, and automation ", Springer-Verlag London Limited, 2005.

[35]   S. Pezeshk, M. ASCE and C. V. Camp, " State of the art on the use of genetic algorithms in design of steel structures ", The university of Memphis, TN, 2000.

[36]   Salwan Samir Sabri, " Optimal Fuzzy Controller Design For (Cúk) Converter Circuit Using Genetic Algorithm ", M.Sc. thesis, college of engineering, university of Mosul, 2008.

[37]   David Beasley, David R. Bull and Ralph R. Martin, " An overview of genetic algorithms : part 1, fundamentals " Inter-university committee on computing, 2005.

[38]   Jarmo T. Alander, " Genetic algorithms : an introduction ", 9th Scandinavian conference on artificial intelligence, Espoo, Finland, 2009.

[39]   Wikipedia, " Crossover (genetic algorithm) ", Wikimedia foundation, Inc. 2011. available at : http://en.wikipedia.org/wiki/Crossover (genetic_algorithm) .

[40]   Wikipedia, " Mutation (genetic algorithm) ", Wikimedia foundation, Inc. 2011. available at : http://en.wikipedia.org/wiki/Mutation (genetic_algorithm) .

[41]   Mithun K., " An exordium of genetic algorithms (GAs) ", Mtech technology management, 2011.

[42]   A. H. Ahmad, A. A. Abdulqader," Power system stabilizer design using real-coded genetic algorithm", 2nd International conference on control, instrumentation and automation ICCIA, Shiraz, Iran, December , 2011.

Appendix (A)

A.1 Power system parameters used in linearized power system   model: (all values are in per unit unless otherwise mentioned).

On a base of 2220 MVA, 24 Kv.:

$F = 60$ Hz	$T_{do}' = 8.0$ sec.
$X_d = 1.81$	$K_D = 0$
$X_d' = 0.3$	$H = 3.5$ sec.
$X_d = 1.76$	$A_{sat} = 0.031$
$X_l = 0.16$	$B_{sat} = 6.93$
$X_E = 0.65$	$\Psi_{TI} = 0.8$
$R_a = 0.003$	$T_R = 0.02$ sec.
$R_E = 0$	$K_A = 200$
$R_{fd} = 0.0006$	$K_{PSS} = 9.5$
$L_{fd} = 0.153$	$T_w = 10$ sec

Nominal operating point is $P_e = 0.9$, $Q_e = 0.3$, $E_t = 1.0$

A.2 Power system parameters used in simpower model, in addition to those given in (A.1)

- Synchronous machine parameters :

$P_n(VA) = 2220e6 \quad V_n(V_{rms}) = 24e3$
$\qquad f_n(Hz) = 60$

$\bar{\bar{X}}_d = 0.23$ p.u. $\qquad\qquad \bar{\bar{X}}_q = 0.25$ p.u.
$\bar{\bar{T}}_{do} = 0.03$ sec.

$\bar{\bar{T}}_{qo} = 0.07$ sec. $\qquad R_s = 0.0025$ p.u. $\qquad p =$
2

- Three-phase power transformer parameters :

$P_n(VA) = 2500e6 \qquad f_n(Hz) = 60$

$V_{1\,ph-ph}(Vrms) = 24x10^3 \qquad R_1 = 1x10^{-6}$ p.u.
$L_1 = 0$ p.u.

$V_{2\,ph-ph}(Vrms) = 230x10^3 \qquad R_2 = 1x10^{-6}$ p.u.
$L_2 = 0.15$ p.u.

- Transmission line parameters : (for 30 km length)

R (Ohms/km) = 0.005 L (H/km) = 0.00129 C (F/km) =
$8.9x10^{-9}$

- Three-phase parallel load parameters :

$P\ (W) = 10x10^6$
$Q_L(\text{positive var}) = 10 \times 10^5$

Appendix (B)

Genetic algorithm execution and fitness function evaluation program written in m-file MATLAB environment.

```
clc
clear
rand('state',0)
tic
  NUM_TRAITS=4;
  HIGHTRAIT=[1.0000 1.0000 1.0000
1.0000];
  LOWTRAIT =[0.1000 0.1000 0.1000
0.1000];
  SIG_FIGS=[5 5 5 5 ]';
  DECIMAL=[1 1 1 1];
  MUTAT_PROB=0.01;
  CROSS_PROB=0.95;
  POP_SIZE=75;
  ELITISM=1;
  DELTA=80;

  EPSILON = 0.001;
  MAX_GENERATION=1000;

%%%% transient performance optimization
specifications
  zeta_ini=0.05;
%%%%%%%%%%%%%%%%%%%%%%%%%%%%%%%%%%%%%%%%%%%%%
%%%%%%%%%%%%%%%%%%%%%%%%%%%%%%%%%%%%%%%
delta_alpha=0.50;dlta=0.50;
delta_zeta=0.05; dltz=0.05;
for zeta_ind=1:10,
    current_zeta=zeta_ini+delta_zeta;
```

```
    delta_alpha=0.50;dlta=0.50;
    alpha_ini=0.00;
    for alpha_ind=1:10,
      current_alpha=alpha_ini-
delta_alpha;
      clear trait bestfitness
bestindividual avefitness k1 k2 k3 k4
k51
      clear k61 k52 k62 k6
rand('state',0)%%%%%%%%%%%%%%%%%%%%%%%%%
%%%%%%%%%%%%%%%%%%%%%%%%%%%%%%%%%%
  popcount=1;
  for pop_member = 1:POP_SIZE
    for current_trait = 1:NUM_TRAITS,

trait(current_trait,pop_member,popcount
)=...
        (rand-
(1/2))*(HIGHTRAIT(current_trait)-
LOWTRAIT(current_trait))+...

(1/2)*(HIGHTRAIT(current_trait)+LOWTRAI
T(current_trait));
      end
    end

CHROM_LENGTH=sum(SIG_FIGS)+NUM_TRAITS;
   TRAIT_START(1)=1;
   for current_trait=1:NUM_TRAITS,
TRAIT_START(current_trait+1)=...

TRAIT_START(current_trait)+SIG_FIGS(cur
rent_trait)+1;
   end
```

```
  while popcount <= MAX_GENERATION
for pop_member = 1:POP_SIZE
for current_trait = 1:NUM_TRAITS,
        if
trait(current_trait,pop_member,popcount
)>HIGHTRAIT(current_trait)
trait(current_trait,pop_member,popcount
)=HIGHTRAIT(current_trait);
        elseif
trait(current_trait,pop_member,popcount
)<LOWTRAIT(current_trait)
trait(current_trait,pop_member,popcount
)=LOWTRAIT(current_trait);
end
%%%%%%%%%%%%%%%%%%%%%%%%%%%%%%%%%%%%%%%%%%%
%%%%%%%%%%%%%%%%%%%%%%%%%%%%%%%
                if
trait(current_trait,pop_member,popcount
) < 0

pop(TRAIT_START(current_trait),pop_memb
er)=0;
                else

pop(TRAIT_START(current_trait),pop_memb
er)=9;
                end

temp_trait(current_trait,pop_member)=..
.

abs(trait(current_trait,pop_member,popc
ount));
```

```
temp_trait(current_trait,pop_member)=...
.
temp_trait(current_trait,pop_member)/10
^(DECIMAL(current_trait)-1);
for      make_gene =
TRAIT_START(current_trait)+1:TRAIT_STAR
T(current_trait+1)-1,
pop(make_gene,pop_member)=temp_trait(cu
rrent_trait,pop_member)-...
rem(temp_trait(current_trait,pop_member
),1);

temp_trait(current_trait,pop_member)=...
.

(temp_trait(current_trait,pop_member)-
pop(make_gene,pop_member))*10;
end
end              end
sumfitness = 0;
%%%%% performing closed loop time
domain analysis as fitness %%%%% %%%%%
calculation method
%%%%%%%%%%%%%%%%%%%%%%%%%%%%%%%%%%%%%%%%%
%
for   chrom_number=1:POP_SIZE,
    clear t1 t2 t3 t4;
  t1=trait(1,chrom_number,popcount);
  t2=trait(2,chrom_number,popcount);
  t3=trait(3,chrom_number,popcount);
  t4=trait(4,chrom_number,popcount);
  f_o=60;
  omega_o=2*pi*f_o;
  xd=1.81;
```

```
xdp=0.3;
xq=1.76;
xl=0.16;
ra=0.003;
rfd=0.0006;
lfd=0.153;
t_dop=8.0;
h=3.5;
kd=0;
a_sat=0.031;
b_sat=6.93;
psi_tl=0.8;
% sensor/AVR parameters %
tr=0.02;
ka=200;
% Network parameters %
re=0;
xe=0.65;
%calculate other parameters %
ladu=xd-xl;
laqu=xq-xl;
ll=xl;
%%%%-- Pss parameters -------- %
kc=9.5;                                  %
tw=10;                                   %
% Set Operating Conditions %
pt=0.9;
qt=0.3;
et=1.0;
% Computing initial steady-state values
of system variables %
it=(sqrt((pt^2)+(qt^2)))/(et);
phi=acos(pt/(it*et));
phi_deg=phi*180/pi;
```

```
% Accounting for saturation effect %
%1-- total saturation effect %
  rho=it;
  [xx,yy]=pol2cart(rho,-phi);
  i_t=(xx)+(yy)*i;
  e_t=et+0*i;
  e_a=e_t+(ra+xl*i)*i_t;

psi_at=sqrt((real(e_a)^2+imag(e_a)^2));
  psi_i=a_sat*exp(b_sat*(psi_at-
psi_tl));
  k_sd=psi_at/(psi_at+psi_i);
k_sq=k_sd;
  xds=k_sd*ladu+ll; lds=xds;
  xqs=k_sq*laqu+ll; lqs=xqs;
  lads=k_sd*ladu;
  lads_p=1/((1/lads)+(1/lfd));
  laqs=k_sq*laqu;
%2-- incremental saturation effect %

k_sd_inc=1/(1+((b_sat*a_sat)*exp(b_sat*
(psi_at-psi_tl))));

lads_inc=k_sd_inc*ladu;k_sq_inc=k_sd_in
c;
  laqs_inc=k_sq_inc*laqu;
  lads_p_inc=1/((1/lads_inc)+(1/lfd));
% calculte delta_internal %
  num=(it*xqs*cos(phi))-
(it*ra*sin(phi));

den=et+(it*ra*cos(phi))+(it*xqs*sin(phi
));
  delta_i=(atan(num/den));
```

```
  delta_ideg=delta_i*180/pi;
% calculate steady-state d&q values %
  eqo=et*cos(delta_i);
  edo=et*sin(delta_i);
  ido=it*sin(delta_i+phi);
  iqo=it*cos(delta_i+phi);
  ebdo=edo-(re*ido)+(xe*iqo);
  ebqo=eqo-(re*iqo)-(xe*ido);
  delta_o=atan(ebdo/ebqo);
  eb=sqrt((ebdo^2)+(ebqo^2));
  ifdo=(eqo+(ra*iqo)+(lds*ido))/lads;
  efdo=ladu*ifdo;
  psi_ado=lads*(-1*ido+ifdo);
  psi_aqo=(-1*laqs*iqo);
% find num/den constants %
  rt=ra+re;
  x_tq=xe+(laqs_inc+ll);
  x_td=xe+(lads_p_inc+ll);
  d=(rt^2)+(x_tq*x_td);
% find m/n constants %
  m1=(eb*((x_tq*sin(delta_o))-
(rt*cos(delta_o)))/d);

n1=(eb*((rt*sin(delta_o))+(x_td*cos(del
ta_o)))/d);

m2=(x_tq/d)*(lads_inc/(lads_inc+lfd));
  n2=(rt/d)*(lads_inc/(lads_inc+lfd));
% calculate the k constants
  k1=(n1*(psi_ado+(laqs_inc*ido)))-
(m1*(psi_aqo+(lads_p_inc*iqo)));
  k2=(n2*(psi_ado+(laqs_inc*ido)))-
(m2*(psi_aqo+(lads_p_inc*iqo)))+((lads_
p_inc/lfd)*iqo);
```

```
   k51=(edo/et)*((-
1*ra*m1)+(ll*n1)+(laqs_inc*n1));
   k52=(eqo/et)*((-1*ra*n1)-(ll*m1)-
(lads_p_inc*m1));
   k5=k51+k52;
   k61=(edo/et)*((-
1*ra*m2)+(ll*n2)*(laqs_inc*n2));
   k62=(eqo/et)*((-1*ra*n2)-
(ll*m2)+(lads_p_inc*((1/lfd)-m2)));
   k6=k61+k62;
% formulate the state matrix a & input
matrix b %
   b=zeros(7,2);
   b(1,1)=1/(2*h);
   b(4,2)=(omega_o*rfd)/ladu;
   a=zeros(7);
   a(1,1)=kd/(2*h);
   a(1,2)=-1*(k1/(2*h));
   a(1,3)=-1*(k2/(2*h));
   a(2,1)=omega_o;
   a(3,2)=-
1*((omega_o*rfd)*m1*lads_p_inc/lfd);
   a(3,3)=-1*(((omega_o*rfd)*((1)-
(lads_p_inc/lfd)+(m2*lads_p_inc))/lfd))
;
   a(3,4)=-1*(b(4,2)*ka);
   a(3,7)=-a(3,4);        %--Pss related
value --%
   a(4,2)=k5/tr;
   a(4,3)=k6/tr;
   a(4,4)=-1/tr;
   a(5,1)=kc*a(1,1);
   a(5,2)=kc*a(1,2);
   a(5,3)=kc*a(1,3);
```

```
a(5,5)=-1/tw;
a(6,1)=a(5,1)*(t1/t3);
a(6,2)=a(5,2)*(t1/t3);
a(6,3)=a(5,3)*(t1/t3);
a(6,5)=(a(5,5)*(t1/t3))+(1/t3);
a(6,6)=-1/t3;
a(7,1)=a(6,1)*(t2/t4);
a(7,2)=a(6,2)*(t2/t4);
a(7,3)=a(6,3)*(t2/t4);
a(7,5)=a(6,5)*(t2/t4);
a(7,6)=(a(6,6)*(t2/t4))+(1/t4);
a(7,7)=-1/t4;
b1=zeros(7,1);
b1(1,1)=1/(2*h);
b1(5,1)=kc/(2*h);
b1(6,1)=kc/(2*h);
b1(7,1)=kc/(2*h);
k3=-1*(b(4,2)/a(3,3));
k4=-1*(a(3,2)/b(4,2));
t_3=-1*(1/a(3,3));
c=[1 0 0 0 0 0 0];
d=0;
ps1s=ss(a,b1,c,d);
[w_natural,zeta_values,P]=damp(ps1s);
cmplx_sort=cplxpair(P);
clear cmplx_pair
for nnn=1:7,
    if imag(cmplx_sort(nnn))~=0,

cmplx_pair(nnn)=cmplx_sort(nnn);
    end
end
alpha_values=real(cmplx_pair);
for alpha_count=1:length(alpha_values),
```

```
      if
alpha_values(alpha_count)>current_alpha
,

chromosome_alpha(chrom_number)=0;break
      else

chromosome_alpha(chrom_number)=1 ;
      end
end
%%%%% now check for Zeta Values to be
in range %%%%%%%%%%%%
if chromosome_alpha(chrom_number)==1,
  for zeta_count=1:7,
        if
zeta_values(zeta_count)<current_zeta,

chromosome_zeta(chrom_number)=0;break
        else

chromosome_zeta(chrom_number)=1;
        end
  end
%%%%%%%    %%%%    %%%%
else %%%% performance index IAE
calculation %%%%%%%%%%
  time1 =0:0.1:15;
  [out,tsamp] = step(ps1s,time1);
  count=0;
  nu_samp=numel(out);
  for ind=1:nu_samp
   count=count+abs(out(ind));
  end
  PA=(count/nu_samp)*100;
```

```
    fitness_bar(chrom_number)=PA;
end
end
for chrom_number=1:POP_SIZE,

fitness(chrom_number)=(1/(fitness_bar(c
hrom_number)));

sumfitness=sumfitness+fitness(chrom_num
ber);
end
[bestfitness(popcount),bestmember]=max(
fitness);
bestindividual(:,popcount)=trait(:,best
member,popcount);
for pop_member = 1:POP_SIZE,
    if ELITISM ==1 &&
pop_member==bestmember
        parent_chrom(:,pop_member)=po
    p(:,pop_member);
    else
      pointer=rand*sumfitness;
      member_count=1;
      total=fitness(1);
      while total < pointer
        member_count=member_count+1;

total=total+fitness(member_count);
      end
parent_chrom(:,pop_member)=pop(:,member
_count);
   end
end
for parent_number1 = 1:POP_SIZE,
```

```
    if ELITISM ==1 &&
parent_number1==bestmember

child(:,parent_number1)=parent_chrom(:,
parent_number1);
    else
       parent_number2=parent_number1;
       while parent_number2 ==
parent_number1
         parent_number2 =
rand*POP_SIZE;
         parent_number2 =
parent_number2-rem(parent_number2,1)+1;
       end
    if CROSS_PROB > rand
         site = rand*CHROM_LENGTH;
         site = site-rem(site,1)+1;
child(1:site,parent_number1)=parent_chr
om(1:site,parent_number1);
child(site+1:CHROM_LENGTH,parent_number
1)=...

parent_chrom(site+1:CHROM_LENGTH,parent
_number2);
    else
child(:,parent_number1)=parent_chrom(:,
parent_number1);
    end
for pop_member= 1:POP_SIZE,
    if ELITISM ==1 &&
pop_member==bestmember
child(:,pop_member)=child(:,pop_member)
;
    else
```

```
      for site = 1:CHROM_LENGTH,
          if MUTAT_PROB > rand
              rand_gene=rand*10;
              while
child(site,pop_member) == rand_gene-
rem(rand_gene,1),
                  rand_gene=rand*10;
              end;

child(site,pop_member)=rand_gene-
rem(rand_gene,1);
              if rand_gene == 10
                    site=site-1;
              end
            end
          end
        end
    end
  pop=child;
  popcount=popcount+1
for pop_member = 1:POP_SIZE
    for current_trait = 1:NUM_TRAITS,
trait(current_trait,pop_member,popcount
)=0;
for
gene=TRAIT_START(current_trait)+1:TRAIT
_START(current_trait+1)-1,
 place=DECIMAL(current_trait)-
place_pointer;
trait(current_trait,pop_member,popcount
)=...

trait(current_trait,pop_member,popcount
)+...
```

```
      (pop(gene,pop_member))*10^place;
 place_pointer=place_pointer+1;
end
if
pop(TRAIT_START(current_trait),pop_memb
er) < 5

trait(current_trait,pop_member,popcount
)=...
        -
trait(current_trait,pop_member,popcount
);
end
    end
end
   if popcount > DELTA+1 && ...
       max(abs(bestfitness(popcount-
DELTA:popcount-1)-...
          bestfitness(popcount-DELTA-
1:popcount-2)))<=EPSILON
       break;
   end
end
[ggg,vvvv]=max(bestfitness);
  t1=bestindividual(1,vvvv);
  t2=bestindividual(2,vvvv);
  t3=bestindividual(3,vvvv);
  t4=bestindividual(4,vvvv);
current_PA=ggg;
    current_err=1/ggg;
PA_mat(alpha_ind,zeta_ind)=current_PA;

err_mat(alpha_ind,zeta_ind)=current_err
;
```

```
    curr_alpha(alpha_ind)=current_alpha;
     delta_alpha=delta_alpha+dlta;
     end
   curr_zeta(zeta_ind)=current_zeta;
   delta_zeta=delta_zeta+dltz;
end
PA_mat_disp=zeros(11);
PA_mat_disp(2:11,2:11)=PA_mat;
PA_mat_disp(2:11,1)=curr_alpha;
PA_mat_disp(1,2:11)=curr_zeta;
toc
```

United Scholars
Publications

Email: info@unitedscholars.net
www.unitedscholars.net

www.ingramcontent.com/pod-product-compliance
Lightning Source LLC
Chambersburg PA
CBHW061321220326
41599CB00026B/4983